A Materials Science Guide to Superconductors and How to Make Them Super

T0266843

A Materials Science Guide to Superconductors and How to Make Them Super

Susannah Speller

University of Oxford

OXFORD
UNIVERSITY PRESS

OXFORD
UNIVERSITY PRESS

Great Clarendon Street, Oxford, OX2 6DP,
United Kingdom

Oxford University Press is a department of the University of Oxford.
It furthers the University's objective of excellence in research, scholarship,
and education by publishing worldwide. Oxford is a registered trade mark of
Oxford University Press in the UK and in certain other countries

Published in the United States of America by Oxford University Press
198 Madison Avenue, New York, NY 10016, United States of America

British Library Cataloguing in Publication Data

Data available

Library of Congress Control Number: 2022932687

ISBN 978–0–19–285834–4 (hbk)
ISBN 978–0–19–285835–1 (pbk)

DOI: 10.1093/oso/9780192858344.001.0001

Printed and bound by
CPI Group (UK) Ltd, Croydon, CR0 4YY

To my wonderful family
who no longer have a good excuse
for not knowing anything about superconductors!

Preface

Superconductors. Even their name evokes the sense that there is something rather special about these materials. But in this day and age when everything is apparently *awesome, amazing* or *incredible*, we have become somewhat desensitised to hyperbolic language. The notion of a conductor whose name implies that is better than other conductors may not seem to be something to get particularly excited about, but these understated materials more than live up to their name. They really are awesome. They are not just a bit better than other materials at carrying electricity, they are such perfect conductors that they carry electricity without losing any energy at all. So why don't we fix the world's energy crisis by replacing all our power cables with superconducting ones? Why don't we use superconductors in all of our power-hungry technology? The fly in the ointment is that superconductors, even the 'high temperature' variety, only work at low temperatures, well below room temperature. This, in itself, is not a huge problem for a lot of applications because cooling is relatively easy nowadays. The real difficulty is that the best superconductors are extraordinarily difficult materials to work with. For a start, they are brittle ceramics. Imagine trying to make a kilometre-long wire from something with the properties of your china mug! They are also chemically very sensitive, even reacting to the moisture in air. And to top it all off, we have to make them essentially as one long single crystal. The fact that we *can* make wires from them and we *can* get those wires to carry extraordinarily high currents is a triumph of materials science over adversity!

This book not only introduces the remarkable physical phenomena that accompany superconductivity, but unlike other books on this topic, it focuses on how we overcome intractable materials properties to enable their powers to be harnessed in real devices. Readers will discover how diverse superconducting materials and their applications are, from the metallic alloys used in the Large Hadron Collider to the thin film superconductors that will be crucial for quantum computers. The book tells about how even the simplest superconductors have to be carefully designed and engineered on the nanometre scale to optimise their performance. Along the way, the reader will be introduced to what materials science is all about and why advanced materials have such widespread importance for technological progress. As it happens, superconductors are an excellent class of material for showcasing the discipline of materials science for various reasons. Firstly, their properties are truly remarkable. Who can fail to be captivated by the magic of levitation and the seemingly impossible notion of frictionless electricity? Their appeal is widespread because not only do they enable 'cool' futuristic technologies (pardon the pun) like hoverboards and quantum computers, but they have the potential to contribute to a greener society. Moreover, superconductors are an amazingly diverse class of materials. They range from simple elemental metals and metal alloys to very complex ceramics containing five or more different elements. Superconductivity has even been discovered in organic molecules and nanomaterials such as derivatives of graphene, and you would be hard pushed to find an element in the periodic table that has not been a constituent in some superconductor or other. They provide the perfect lens through which to examine how to manufacture technological materials which have to be carefully tailored on the nanoscale to optimise performance. The tricks that are used to make superconductors carry very large currents are the same ones that are used to make, say, high performance structural steels or nickel superalloy turbine blades, and these are themes that are explored in the book.

The approach taken in the book is rather different to other 'popular science' books on materials. Instead of introducing a range of different 'quirky' materials, this book aims to convey the essence of materials science in a fun and engaging way by focusing primarily on superconductors, which are a fabulous playground for materials scientists. The book communicates the central theme of materials science by painting a picture of the importance of controlling everything from atomic scale chemistry and bonding right up to the macroscopic scale of large machines and devices. It is also very different from other (very good) books on superconductivity for a general audience, which typically focus on the physics of superconductors and the historical developments, rather than on how we make these materials in practice.

No prior knowledge of materials science or superconductors is needed to enjoy this book! It is written as a narrative using simple language and explaining scientific concepts in an accessible way. The story can be read on its own, but an interested reader can choose to delve deeper into some of the maths and science in additional *Under the Lens* sections. These are standalone text boxes and appendices that accompany the narrative, and go a step further to give readers wishing to stretch themselves an opportunity to delve deeper into the science and see how some simple mathematics can be used to estimate key properties. The *Wider View* sections show how the same things that we care about in superconductors are important in other materials, from the superalloys used at extreme temperatures in jet engines, to the transparent conducting oxide films that are used in the touch screens of smart phones and tablet computers.

It is a great pleasure to thank the people who have helped during the conception and writing of this book. I am grateful for the support of Sönke Adlung and his team at Oxford University Press, and the encouragement of Stephen Blundell that spurred me forward with this project. I would also like to thank my colleagues in the Department of Materials and St Catherine's College for providing a stimulating and friendly working environment. A special mention goes to Chris Grovenor, who has not only been a fantastic mentor and friend for over twenty years, but has also read the entire manuscript and helped me improve it by covering it in his usual red pen!

Finally, I would like to dedicate this book to my loving and long-suffering family, without whose unwavering support I would have long since given up on my dream of having an academic career, and who no longer have a good excuse for not knowing anything about superconductors!

Contents

1

Meet the Family

Frictionless electricity and levitation: you would be forgiven for thinking that these belong in the realms of science fiction or the magical world of Harry Potter, but with superconductors these things are a reality. Equally amazing is the fact that we can actually manufacture kilometres of resistance-free superconducting wire containing hundreds or thousands of continuous copper-clad superconducting filaments, each less than the diameter of a human hair. What is more, we can even make these wires from superconductors that have the mechanical properties of a teacup. This book tells the remarkable story of how it is done. Along the way, we will see lots of aspects of materials science—the science of technology of how to make things—in action, from understanding how the performance of the materials are affected by their chemistry and microscopic internal structure, to finding out how to make the materials to get the most out of their performance.

1.1   Basic properties of superconductors

Superconductors are aptly named because they really are *super* electrical conductors. In fact they conduct electricity so well that no energy is lost in the process at all. They have absolutely zero electrical resistance. This has the major benefit of saving energy, but it also means that superconducting wires can carry much more current than the same sized wire made from a conventional metal which is vital for making compact magnets. To understand why, it is helpful to think about what happens in a circuit made from a normal metallic conductor like copper. You know from experience that a power supply such as a battery is needed to keep the current flowing around the circuit. For example, a torch will stop working when its battery runs out. The battery provides a force that starts electrons in the metal moving along the wire. Newton's second law of motion tells us that a force exerted on an object will make it accelerate. This means that electrons should get faster and faster under the influence of the electromotive force (voltage) provided by the battery, but every so often they bump into things and get scattered—knocked off course—and lose some of their forward momentum. Therefore, if the power supply is switched off, instead of keeping on going at the same speed, the electrons rapidly slow down and stop. (They do not actually stop moving altogether, but they move in random directions and the average flow of charge along the wire becomes zero.) This is exactly the same thing that happens if you give a shopping trolley a push—it won't get very far before the resistance due to friction brings it to a standstill.

Fig. 1.1: Photograph of a superconductor levitating above a magnet. (Courtesy of University of Cambridge.)

Now think what would happen if electrons could flow with absolutely no resistance to their motion. Once you have started them moving around a closed circuit, you can disconnect the power supply. It is not needed anymore. There is nothing to slow the electrons down and so the current carries on forever! This is the equivalent of having a car with no friction or air resistance. You would only need the engine to get it moving in the first place and then it could be turned off and the car would carry on moving at the same speed in the same direction until it runs out of road. Sounds impossible, but that is exactly what Newton's first law of motion tells us—an object with no resulting force acting on it will keep going at the same velocity—and it really happens with the electrons in a superconductor. We have perfectly *frictionless electricity*.

Another captivating and slightly spooky property of superconductors is that they can levitate in mid-air above a magnet. You may have seen videos of this effect in action, with the vapour from the liquid nitrogen used to cool the superconductor adding to the other-worldly, magical effect (Fig. 1.1). You may even have seen sumo wrestlers being levitated on platforms or wacky hoverboards floating gracefully above the ground. These effects happen because of the unique way that superconductors react if you put them in a magnetic field. They push the magnetic field away rather than letting it enter the material, producing a magnetic repulsion force. This expelling (pushing out) of magnetic flux is known as the *Meissner effect*. Even more weirdly, in practice it is possible to trap magnetic field lines inside the right kind of superconductor to achieve a much more stable levitation. The superconductor essentially becomes locked in place relative to the magnetic field, making it possible to actually suspend it in mid-air beneath a magnet or make it whizz around a magnetic track following the field around the corners without needing to be steered.

Fig. 1.2 legend:
- Superconducts above 4.2 K
- Superconducts below 4.2 K
- Superconducts under pressure
- Does not superconduct

1	2	3	4	5	6	7	8	9	10	11	12	13	14	15	16	17	18
H																	He
Li	Be 0.023 K											B	C	N	O	F	Ne
Na	Mg											Al 1.2 K	Si	P	S	Cl	Ar
K	Ca	Sc	Ti 0.4 K	V 5.4 K	Cr 3.0 K	Mn	Fe	Co	Ni	Cu	Zn 0.85 K	Ga 1.1 K	Ge	As	Se	Br	Kr
Rb	Sr	Y	Zr 0.61 K	Nb 9.3 K	Mo 0.92 K	Tc 7.8 K	Ru 0.49 K	Rh 0.0003 K	Pd 3.3 K	Ag	Cd 0.52 K	In 3.4 K	Sn 3.7 K	Sb	Te	I	Xe
Cs	Ba	La* 4.9 K	Hf 0.13 K	Ta 4.5 K	W 0.015 K	Re 1.7 K	Os 0.66 K	Ir 0.11 K	Pt 0.019 K	Au	Hg 4.2 K	Tl 2.4 K	Pb 7.2 K	Bi	Po	At	Rn
Fr	Ra	Ac**	Rf	Db	Sg	Bb	Hs	Mt	Dm	Ry	Uub						

* Lanthanides:

58	59	60	61	62	63	64	65	66	67	68	69	70	71
Ce	Pr	Nd	Pm	Sm	Eu	Gd	Tb	Dy	Ho	Er	Tm	Yb	Lu

** Actinides:

90	91	92	93	94	95	96	97	98	99	100	101	102	103
Th 1.4 K	Pa 1.4 K	U 0.2 K	Np	Pu	Am 0.6 K	Cm	Bk	Cf	Es	Fm	Md	No	Lr

Fig. 1.2: Periodic table showing prevalence of known superconductors. Superconducting transition temperatures are indicated below the chemical symbols of the superconducting elements.

1.2 Superconducting materials

The properties of superconductors are so unusual that you may think that superconductivity must be an incredibly rare phenomenon reserved for a very special kind of material, but this could not be further from the truth. There is an absolutely vast array of superconducting materials. Superconductivity has been found in materials as diverse as elemental metals, complex oxide ceramics, nanomaterials and recently even hydrides that are gases under ambient conditions. Back in the 1970s, Bernd Matthias looked at the occurrence of superconductivity in elements across the periodic table, and even then he commented (in German) that: 'Almost everything is superconducting.' An adapted version of his periodic table is shown in Fig. 1.2. Even the elements that are not superconducting (or only superconduct under pressure) on their own are often key components in superconducting compounds. Notable examples are copper and iron, each of which has its own entire family of superconducting compounds. In fact, we are hard pushed to find any elements at all that do not feature in some superconductor or another.

So let's go back to the start of the story. The phenomenon of superconductivity was discovered originally by Heike Kamerlingh Onnes in 1911. Along with various other scientists of the day, he was working on the low temperature electrical properties of metals, in particular trying to find out what happens to the resistance of high purity elements as the temperature approaches absolute zero. Absolute zero is the lowest possible temperature, the temperature at which a material's thermal energy store is zero. It has a value of −273.15°C, but it is often convenient to use the absolute temperature scale (measured in units of kelvin (K) where the temperature in K is just the temperature in °C shifted by 273.15. This means that absolute zero is 0 K, water freezes at 273.15 K and water boils at 373.15 K. Kamerlingh Onnes managed to be the first person to liquefy helium, which meant he could access colder temperatures than anyone else. In his original experiment, he measured the electrical resistance of the purest metal he could make, which happened to be mercury. He found an extraordinary thing: instead of its resistance gradually decreasing as the temperature reduced (as one popular theory of the day predicted), or starting to increase again (as another theory suggested), at around 4 K its resistance suddenly dropped to an unmeasurably low value. When we say *unmeasurably low*, the sensitivity of his equipment was around a millionth of an ohm (Ω). Subsequently various experiments have been performed to try to get more accurate estimates. These essentially involve starting a current going in a superconducting ring and measuring for a very long time to see how fast the current decays. From these experiments we know that the resistivity is lower than 10^{-26} Ω m (0.000...0001 with 26 zeros before the 1), but nobody has counted for long enough to see any measurable decay yet!

Under the Lens

Orders of magnitude

We are going to come across some very small numbers and some very big numbers in this book. It is very inconvenient to have to write them out in full, so scientists and mathematicians use a notation called standard form, or special prefixes for units like *kilo*. Standard form is based on powers of 10. For example 10^2 is $10 \times 10 = 100$ and $10^3 = 10 \times 10 \times 10 = 1000$. For small numbers we use negative indices. $10^{-1} = \frac{1}{10} = 0.1$ and $10^{-2} = \frac{1}{10^2} = 0.01$. The main prefixes that we will come across are given below.

Table 1.1: List of unit prefixes.

Prefix	Symbol	Standard form	Number
tera	T	10^{12}	1,000,000,000,000
giga	G	10^{9}	1,000,000,000
mega	M	10^{6}	1,000,000
kilo	k	10^{3}	1,000
		10^{1}	1
milli	m	10^{-3}	0.001
micro	μ	10^{-6}	0.000001
nano	n	10^{-9}	0.000000001
pico	p	10^{-12}	0.000000000001

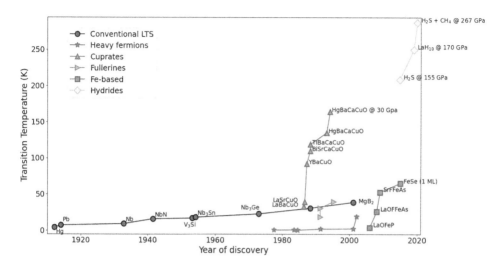

Fig. 1.3: Historical discoveries of superconducting materials.

As a brief aside, it is worth mentioning the difference between *resistance* and *resistivity*. The eagle-eyed of you may have noticed that they have different units. Resistance is defined as the voltage required to keep 1 amp of current flowing in a wire, and it will depend on the length and thickness of the wire as well as what material the wire is made of and other things like temperature. This is rather inconvenient if we want to compare the properties of different materials. We would much rather work with a *materials property* that does not depend on the physical size or geometry of the wire. That is where resistivity comes in. Since resistance scales with length of the wire, dividing the resistance by the length of the wire gives us a way of comparing wires of different lengths—its resistance per unit length. But the thicker the wire, the easier it is for the current to flow, so the lower the resistance. (Think of how much harder you have to suck to drink a milkshake through a narrow drinking straw compared to a wide one.) This means that we also have to multiply the resistance by the cross-sectional area of the wire to get a property that does not depend on wire thickness. Putting these together, we define resistivity as $\rho = \frac{RA}{L}$, where R is the resistance of the wire, A is its cross-sectional area and L is its length.

Since the initial discovery of superconductivity, many thousands of materials which show this same electrical phenomenon have been discovered, a small fraction of which are shown in Fig. 1.3. Initially work focused on pure metallic elements which typically only superconduct at extremely low temperatures, but over the following decades it was discovered that high purity is not a requirement for superconductivity. Surprisingly, metal alloys were found to have superior superconducting properties. But it was not until the late 1980s that the next major breakthrough happened—the discovery of *high temperature superconductors*. Do not get too carried away though. High temperature in this context does not mean hundreds of degrees Celsius. It does not even mean room temperature. What it means is that instead of needing less than −250°C (about

20 K), they still superconduct at temperatures above $-190°C$ (about 80 K). This may not sound like a big deal, but in practice it is a lot easier and cheaper because instead of needing liquid helium with a boiling point of 4 K, we can use liquid nitrogen which has a boiling point of 77 K. Liquid nitrogen costs about the same as milk (whereas liquid helium costs about the same as whisky!), so this opens up a much wider range of mainstream applications for superconductors. The most obvious of these is probably the idea of replacing the long distance overhead power lines with a much more energy efficient alternative for getting electricity from A to B. The discovery was surrounded by an enormous media hype. The conference where the discovery was announced was dubbed the 'Woodstock of Physics' by analogy with the seminal Woodstock music festival of 1969. High temperature superconductors were thought to be the breakthrough needed to solve all the world's energy problems.

However, the 'superconductivity revolution' promised by Time magazine (among others) has not happened—yet. So what went wrong? It turned out that, even though these materials have fantastic superconducting properties, they are extraordinarily difficult to make into wires that can carry high currents. The materials themselves are so difficult to work with that it is only now, over 30 years after their discovery, that we are poised to use them in anger. The fact that we can do it at all is a triumph of materials science. These materials are ceramics, so they are inherently brittle. They need high temperature processing but they are incredibly sensitive to precise chemical composition and get damaged by reactions with most other things including metals and water. They also need to be made essentially as one long single crystal in order to carry currents that are high enough to be useful for real applications. The fact that you can go and buy yourself a kilometre long piece of one of these materials, if you can afford the extortionate price tag, is frankly almost a miracle.

The Wider View

Classification of materials

Traditionally materials are grouped into three different classes, based primarily on their chemistry and structure: metals, ceramics and polymers. This makes sense because the properties of materials are largely governed by what atoms they are made of and how those atoms are arranged and bonded to each other.

Table 1.2: Summary of classes of materials.

Class	Bonding and structure	Common properties	Examples
Metal	Giant metallic structure	Hard, ductile, good conductors of heat and electricity	Copper, steel
Ceramic	Giant ionic or covalent structure	Hard, brittle, high melting point, electrical insulators	Alumina (Al_2O_3), quartz (SiO_2)
Polymer	Long chain organic molecules	Soft, flexible, low melting point	Polystyrene, polyethylene (plastics), rubber

Composites are materials that are made up of more than one constituent: a matrix (main phase) and a filler . Carbon fibre reinforced plastic and fibreglass are examples of polymer matrix composites with ceramic fibres to provide strength, but metal matrix and ceramic matrix composites also exist. In general, materials fit quite nicely into one box or another, but there are some exceptions to this and high temperature superconductors are one of them. These materials are based on copper oxide, which ought to be ceramic, and their brittleness is consistent with that. However, unlike conventional ceramics, they conduct electricity even at room temperature and their electrical properties are characteristically metallic.

1.3   Critical parameters

Superconducting materials only superconduct under certain circumstances. The first of these we have already alluded to: they must be kept cold. The temperature at which the resistance of the material drops to zero is called the *superconducting transition temperature* or *critical temperature* (T_c). The second condition that must be met is that they must not be exposed to too high a magnetic field. The maximum magnetic field is called the *critical field* (B_c). Finally, there is a maximum current that can be passed through a given cross-section of the superconductor before it loses its zero resistance properties—the *critical current density* (J_c). These parameters vary from material to material, and in the case of J_c is strongly dependent on how defective the material is on the nanometre and micrometre scale. In complete contrast to semiconductors like silicon that have to be as perfect as possible, the more defects in superconductors the better. This makes them a playground for materials scientists. There are loads of tricks we can play to optimise and tailor their properties, and there is always something interesting to see inside the material when put under a microscope!

The three critical parameters are actually related to one another. For example, the critical field and critical current density both decrease with increasing temperature. This means that the further below T_c you cool the superconductor, the higher the magnetic field and/or current density that the material will tolerate before superconductivity is lost. Similarly, critical current density also decreases with applied magnetic field. These interdependencies are often visualised using a concept called the *critical surface*, as shown in Fig. 1.4 which gives the extent of superconductivity on a three-dimensional graph of temperature, magnetic field and current density. The material will be in the superconducting state for any combination of conditions that are underneath this critical surface (closer to the origin). A lot of the work of materials scientists is to push this envelope outwards, to maximise the practical operating window of the superconductor.

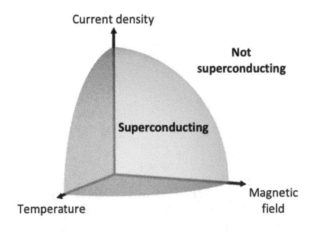

Fig. 1.4: Critical surface of a generic superconductor.

1.4   Forms of superconductor

As well as there being a very diverse range of materials that superconduct, we can also make them in a variety of different forms (Fig. 1.5). For the vast majority of applications, what we want is long lengths of wire for carrying current. In general we call these *conductors*. An individual wire is known as a *strand* and is typically about 1 mm in diameter. Most strands are *multifilamentary*—they consist of hundreds or thousands of separate fine superconducting *filaments* embedded in some sort of metal matrix. These filaments are what carries the current and they must be continuous along the entire length of the strand. For many applications, the most desirable form of conductor is a multifilamentary wire with a circular cross-section—*round wire*. Unfortunately, some high temperature superconducting compounds cannot be processed in this form so manufacturers have to resort to making multifilamentary tapes (which are essentially squashed round wires) or, in the worst case scenario, complicated and expensive multilayered tapes called *coated conductors* in which each layer is laid down in turn as a coating on top of the previous layer. For some applications multiple superconducting strands need to be assembled together into a *cable* before making the device.

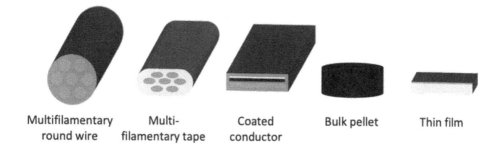

| Multifilamentary round wire | Multi-filamentary tape | Coated conductor | Bulk pellet | Thin film |

Fig. 1.5: Different forms of superconductor.

Conductors are not the only useful form of superconductor though. The impressive magnetic levitation demonstrations that you might have seen use pellets of superconductor known as *bulks*. These are generally cylindrical in shape, although other geometries like rings have also been manufactured for special purposes. Their use is not just limited to showing off some of the weird magnetic properties of superconductors, they can be used in magnetic levitating (maglev) trains and flywheels for energy storage, and as permanent magnets for electrical motors and generators. The final form in which superconductors are commonly used is as *thin films*. These are very thin layers, a few hundreds of nanometres thick, that are grown on top of some sort of non-superconducting substrate. Electronic circuits of various types can be made using these thin films by selectively removing the superconductor from some areas to define tracks that the current can flow along.

1.5   The technological superconductors

Even though there are an incredible number of different superconducting materials to choose from, there are only six different superconductors that are commercially available in the form of long lengths of wire. We call these the *technological superconductors*. They are not necessarily the best materials in terms of raw superconducting properties, but they have emerged as having a workable balance between superconducting performance, processability and (to some extent) cost. Three of the materials are classed as low temperature superconductors: niobium-titanium (NbTi), niobium-tin (Nb_3Sn) and magnesium diboride (MgB_2). The other three are high temperature superconductor (HTS) and they have much more complicated chemical formulae, so we refer to them as Bi-2212 ($Bi_2Sr_2CaCu_2O_8$) and Bi-2223 ($Bi_2Sr_2Ca_2Cu_3O_{10}$), with the numbers reflecting the ratio of the cations in the compound, and (RE)BCO ($REBa_2Cu_3O_7$). The main properties of these six materials are shown in Table 1.3.

Ideally magnet designers want wires that contain a whole load of fine superconducting filaments embedded in copper. This is because the copper does a good job at conducting away heat or acting as an electrical shunt for the current if there is a problem with the superconductor. Engineering critical current density (J_e) is defined as the critical current divided by the whole cross-sectional area of the wire including the copper. As can be seen in Fig. 1.6, J_e values of the low temperature superconducting materials drop off quite steeply with increasing field, with Nb_3Sn outperforming both NbTi and MgB_2. By contrast, the high temperature superconductors still have high J_e values at enormous magnetic fields of 45 tesla. For comparison purposes, the magnetic field of one of those strong permanent magnets you can buy has a field of about 1 tesla. The earth's magnetic field is about 0.00001 (10^{-5}) tesla. It is worth pointing out that Fig. 1.6 has two different lines for (RE)BCO and Bi-2223. This is because these materials are only available in the form of tapes, and their current-carrying properties are much better when the magnetic field is applied parallel to the surface of the tape ($B\|$) than when it is applied perpendicular to the surface of the tape ($B\perp$).

1.6   Applications of superconductors

The potential applications of superconductors are as diverse as the materials themselves. At one end of the spectrum, the largest superconducting machine that currently exists is the Large Hadron Collider (LHC) at CERN, which is used for smashing atoms apart. Superconducting magnets are used all around the 27 km ring to bend the proton beams as well as in the detectors that pick up the subatomic particles. These magnets operate at temperatures of 1.9 K, colder even than the 2.7 K of outer space! Superconducting magnets are also important in nuclear fusion reactors such as ITER[1] that is under construction in France. High strength magnets are essential for confining the fusion reaction that takes place at a temperature hotter than our sun where no material can survive! ITER uses a low temperature superconductor, Nb_3Sn but the next generation compact reactors are likely to use high temperature superconductors.

---

[1]ITER was originally an acronym for 'International Thermonuclear Experimental Reactor', but the full version of the name is no longer used.

Table 1.3 Summary of the six technological superconductors.

Material	Form of wire	T_c	Mechanical properties	Stabiliser
NbTi	Multifilamentary round wire	9 K	Ductile	Copper
Nb$_3$Sn	Multifilamentary round wire	18 K	Brittle	Copper
MgB$_2$	Multifilamentary round wire	39 K	Brittle	Fe or Ni alloy
Bi-2212	Multifilamentary round wire	85 K	Brittle	Silver
Bi-2223	Multifilamentary tape	110 K	Brittle	Silver
(RE)BCO	Coated conductor	93 K	Brittle	Fe or Ni substrate

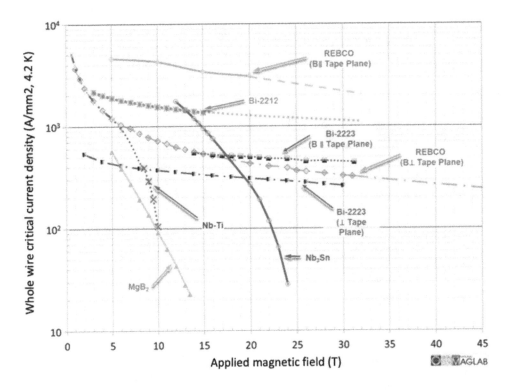

Fig. 1.6: Engineering critical current density as a function of applied magnetic field at 4.2 K for the technological superconductors. The data was compiled by the National High Magnetic Field Laboratory in Florida from various sources: (Braccini *et al.*, 2010; Jiang *et al.*, 2019; Boutboul *et al.*, 2006; Kanithi *et al.*, 2014; Parrell *et al.*, 2003; Parrell *et al.*, 2004; Li *et al.*, 2013).

At the other end of the scale, there are numerous thin film superconducting devices for applications such as incredibly sensitive detectors and quantum computing. Somewhere in the middle are the 'green' power applications—motors and generators, power transmission lines and superconducting magnetic energy storage (SMES)—and levitation applications including frictionless bearings and maglev trains. By far the most important commercial application of superconductors so far is for the big magnets that you get wheeled inside when you have a magnetic resonance imaging scan in hospital and the high-field magnets used for the related nuclear magnetic resonance technique used in labs across the world for characterising materials.

These applications are often categorised as either large scale or small scale. Large scale includes pretty much anything that is made from wires or bulk superconductors, ranging from desktop systems right up to the scale of the LHC, as shown in Fig. 1.7. In the vast majority of these devices, the materials need to be engineered to carry as high a current density as possible under the operating conditions. These range from very high magnetic fields for fusion magnets and nuclear magnetic resonance instruments

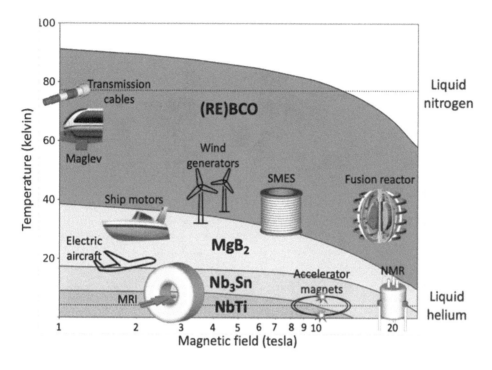

Fig. 1.7: Large scale applications of superconductors, giving an indication of the operating conditions of current and future devices.

to the relatively high temperatures desirable for many power applications. However, the materials requirements of small scale devices are quite different. High currents, high magnetic fields and high temperatures are rarely the main drivers. Instead, these devices tend to need more 'perfect' materials with fewer defects and very high quality surfaces to minimise losses that can compromise device performance.

It is also worth mentioning some interesting applications that rely on the fact that superconductors can be pushed beyond their limits, where they become normal metallic conductors, and then revert back to being superconducting again naturally when the conditions improve. One of these is the *fault current limiter*, which is a superconducting version of a fuse or circuit breaker that can automatically reset itself. When a fault current—one that is unusually high—flows through the superconducting device, the critical current density is exceeded and suddenly some resistance is generated. The current is diverted into a different circuit to protect the system. When the fault has been resolved, the current drops, superconductivity naturally returns and normal operation is restored. Another example is a rather simple small scale thin film device that is used for detecting weak electromagnetic signals from space. The device is essentially a long, meandering track etched into the surface of a superconducting film with a current flowing through it. When light of some sort hits the track, enough energy is dumped into the superconductor for a small hot spot to stop superconducting. The current has to squeeze through a narrower region to get past the hot spot, and that concentration of current is sufficient to exceed the critical current of the entire track. The voltage pulse that is produced when the track temporarily becomes resistive is the signal that is detected. These detectors are very sensitive, can be tuned to different wavelengths of light and also have the advantage of being fast, recovering quickly when superconductivity is naturally re-established.

The main focus of this book is to understand how materials scientists design and engineer superconductors to tailor their properties for specific applications. In the majority of cases, this is all about getting as high a current density as possible through the material. The exception is small scale electronic devices that will be touched upon in Chapters 6, 9 and 10. Before we dive into the specifics of the different classes of superconducting materials, we are going to begin our story by looking at magnets because they are the main application of superconductors to date, and in particular the most successful commercial application of superconductors—magnetic resonance imaging.

2

Miraculous Magnets

Although superconductors have a huge variety of possible applications, from small scale electronics up to long distance power transmission, it is as magnets that they have really come into their own. To date, the vast majority of applications of superconductors revolve around using them as magnets in some shape or form. Part of the reason is that magnets are essential components in a surprising number of different technologies. They are needed to bend beams of charged particles in accelerators like the Large Hadron Collider. They are needed in nuclear fusion reactors to confine the intensely hot plasma—as hot as the sun—where physical materials would not stand a chance of surviving. Magnets are needed in electrical machines like motors and generators for the nuclear magnetic resonance (NMR) instruments used in scientific laboratories, for magnetic separation of minerals, for magnetic energy storage, for magnetic propulsion systems, for magnetic drug targeting therapies, and so on. Not surprisingly, magnets are also an integral part of magnetic resonance imaging (MRI) machines that are now commonplace in hospitals and are used to take images of the insides of our bodies when we injure ourselves—the clue is in the name! For many of these applications, superconductors are either essential or provide significant benefits over non-superconducting alternatives. In this chapter we will explore why superconductors and magnets go hand in hand. We will particularly focus on MRI because this is currently by far the most important commercial application that uses superconductors in anger. In fact, MRI and NMR (its sister technique) account for about 90% of the ~5 billion dollar global superconductivity market, with over 40,000 MRI machines installed worldwide, and without superconductors the technique is simply not viable.

2.1   Why are superconductors used for magnets?

Before we can understand why superconductors are so useful for magnet applications, we need to think about what magnetic fields are and how they are created. A magnetic field is essentially a force field that is produced by a magnet and acts on other magnetic objects. These objects do not themselves have to be permanent magnets; to experience a force they just need to respond in some way to the magnetic field they are placed in. In fact, the term *magnetic field* is not very specific because there are two different quantities that have different units that are both loosely referred to as magnetic fields: the B-field and the H-field. The B-field is more formally called the *magnetic induction* or the *magnetic flux density*, and it is measured in units of tesla (T). For the purposes of this book, this is the only field we need to worry about so we will simply refer to it as the magnetic field. Probably the first thing you think of when someone says the word 'magnet' is a fridge magnet, or possibly one of those bar magnets or horseshoe magnets which have one end painted red to denote the north pole. You may have done experiments with iron filings on paper to look at the pretty patterns produced by the field lines of magnets arranged in different ways. You are probably also aware that navigational compasses work by picking up the Earth's magnetic field. Compass needles are themselves permanent magnets that pivot around the central point to line up with the (magnetic) north pole. In the materials used for making bar magnets and compass needles, it is the atoms themselves (or some of them) that are magnetic. Their magnetism actually originates from the weird quantum physics that governs particles like electrons on the very small scale. For example, an iron atom is magnetic because the particular number and arrangement of its electrons leads to it having its own north and south poles. We call this a dipole (because it has two poles) and the strength of the magnet is called the *dipole moment*.[1] If the magnetic poles of all of the iron atoms are pointing in the same fixed direction, we have a *permanent magnet*—one that generates its own magnetic field. This is completely different from how superconducting magnets work. Although, as you will see in Chapter 4, we can turn chunks of superconductor into permanent magnets that behave rather like very strong fridge magnets, their magnetism does not come about because the individual atoms inside them are magnetic.

Superconducting magnets are actually much more like classical *electromagnets* in which the magnetic field arises as a result of a flow of electrical charge—a current. The fact that electrical currents produce magnetic fields was discovered simultaneously by two famous scientists, Ampère and Oersted, back in 1819. Ampère is now best known because the unit of current, the ampere or amp, was named after him. In fact, Oersted was also honoured by having a unit of magnetic field strength named after him, but although it is still used by some people today, it is not an SI (international standard) unit so it is not as well known.[2] In his experiments, Oersted used a compass to map out the direction of the magnetic field around wires carrying currents. He found that the magnetic field generated by a long straight wire forms a circular ring pattern around the wire. Reversing the direction of the current in the wire was found to flip the direction of the magnetic field (the north pole of the compass needle rotated by 180°), but the shape the field lines produced remained circular. Unsurprisingly, the strength of the magnetic field was found to increase with increasing current. It was also discovered that the field strength drops off pretty quickly with radial distance from the wire, so the closer you are to the wire, the stronger the magnetic field will be. For this reason, long straight wires are not particularly useful for making magnets. However, if you wrap the wire into a loop, the magnetic field gets concentrated into the centre, and you end up with a magnetic field pattern that looks identical to a short bar magnet. If we hid a perfectly circular current loop and a short bar magnet inside

---

[1]You may have come across the term 'moment' before in relation to a turning moment. In that case it is defined as the force multiplied by the distance of the force from the pivot point. In a similar vein, a dipole moment is the strength of the magnetic poles multiplied by the distance between them.

[2]1 oersted is equivalent to $\frac{10^3}{4\pi}$ A m^{-1}

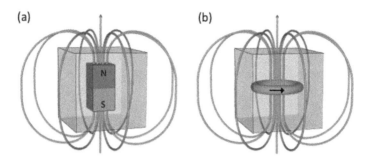

Fig. 2.1: Equivalent magnetic field distributions around (a) a bar magnet, and (b) a current loop.

two identical boxes and measured the magnetic field distribution outside, we would not be able to tell which object was inside which box, as shown in Fig. 2.1.

Often we want to create a more uniform field—one in which all the field lines point in the same direction and where the strength of the field does not vary with location. To do this we can wrap the wire around lots of times to form a long coil—a *solenoid*—that looks a bit like a spring (Fig. 2.2). If the solenoid is long enough, the magnetic field will have the same strength everywhere inside its bore and the field will point straight along its axis. Outside an infinitely long coil, the field will be zero. The field strength inside a solenoid not only depends on how big a current we pass through the wire, but also on how many loops of wire (turns) we manage to squeeze in along its length. The more turns per unit length, the stronger the field will be. This means that to maximise the field strength and keep the magnet compact, we actually need to get as much current as we can through as thin a wire as possible. Consequently, a key property of the conducting material that we choose to use for the wire in a magnet is the maximum *current density* that it can carry. This is just the maximum current that can be safely passed through the wire divided by its cross-sectional area.

So what limits the current density that a normal metal conductor like copper can carry? As a result of its resistance, when a large current is passed through the wire a lot of power is dissipated as thermal energy. This heats up the material and is known as *Joule heating*. The resistance of normal metals increases with temperature

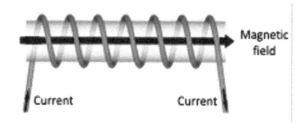

Fig. 2.2: Magnetic field inside a long solenoid.

because there is more chance that the electrons will be knocked off course (scattered) if the atoms in the material are jiggling about more. This means that, unless heat is extracted as quickly as it is generated, we end up with a thermal runaway situation and the metal will eventually melt. A back-of-the-envelope calculation suggests that the maximum current that we can safely pass through copper wire with cross-sectional area of 1 mm^2 is less than 20 A. The magnetic field that can be generated by a solenoid with a single layer of windings made of this wire will be pretty small—around 0.02 tesla. This is not much more than the strength as a normal fridge magnet, and only about a fiftieth of the strength of the strongest permanent magnets that you can buy. We could increase the field by adding more layers of windings to increase the number of turns per unit length, but this quickly leads to the outer dimensions of the magnet getting rather large.

So how do superconductors compare? As we know, superconductors can carry current with zero resistance, so there is no heat generated—provided they stay superconducting. Current is therefore not limited by Joule heating in the same way as it is in conventional conductors. Instead, it is limited by the *critical current density* of the superconductor—the maximum current density that it can sustain without losing its zero resistance property. The critical current density of a superconductor is influenced by lots of different factors. Not surprisingly, it depends what elements the material is made from—the chemical makeup of the material—and how the atoms are arranged and bonded together. Less obvious is the fact that it also depends on how perfect the material is on the nanometre to micrometre length-scale. Counter-intuitively, it turns out that materials that are more defective on the nanoscale typically have higher critical current densities, for reasons that will be discussed in detail in Chapter 4. In addition to the nature of the material itself, the operating conditions of temperature and magnetic field also affect the critical current density. The higher the operation temperature and field experienced by the superconductor, the lower its critical current density will be (Fig. 1.4).

Additionally, it is worth remembering that superconducting wires are never made of 100% superconductor. To reduce their tendency to suddenly stop superconducting if, say, the temperature rises too much at some location, a whole heap of copper is put into the wire to help dissipate heat quickly and carry some of the current if really necessary. As mentioned in Section 1.5, for engineering purposes the critical current density calculated using the entire cross-sectional area of the wire (including the copper components) is more instructive than the critical current density of the superconducting component alone. Typically, for the technological superconductors we can reach engineering (whole wire) current densities of well over 1000 A mm^{-2} at low magnetic fields. Amazingly, some of the high temperature superconductors can still carry in excess of 500 A mm^{-2} at the highest fields we can access to measure them ($\sim$ 40 tesla), provided we cool them to liquid helium temperature (4 K). Obviously, this current density is far in excess of what can be sustained in a conventional metal like copper ($\sim$ 20 A mm^{-2}). This means that we can make superconducting magnets that are 25–50 times stronger than copper magnets of the same geometry. Alternatively, if we do not need extremely high magnetic fields, we can use the enormous critical

Under the Lens

Current rating estimate

For a stable current to flow in a normal metallic conductor like copper, the heat generated per second by the resistance must be able to dissipate to the surroundings (radiate away). Let's take a wire with square cross-section of side length x and length l. If ρ is the resistivity of the material we can work out the power P (energy per second) dissipated by the resistance R from current I.

$$R = \frac{\rho l}{x^2} \tag{2.1}$$

$$P = I^2 R = \frac{I^2 \rho l}{x^2} \tag{2.2}$$

The rate of heat energy Q radiated out of an object scales with its surface area A and with the fourth power of temperature according to Stefan's law (equation 2.3). Since the cross-sectional area of the small ends of the wire can be ignored, $A \approx 4xl$ (for the four long rectangular faces).

$$P = \frac{dQ}{dt} = kAT^4 \approx 4kxlT^4 \tag{2.3}$$

Here k is a materials constant that is related to the *emissivity* of the material (how easily it emits heat).

To find the maximum current for stable operation, we need to look for the condition where the electrical power and the radiated power are equal.

$$\frac{I^2 \rho l}{x^2} = 4kxlT^4 \tag{2.4}$$

$$\frac{I^2 \rho}{x^2} = 4kxT^4 \tag{2.5}$$

As we can see, the equations are not affected by the length of the wire. Let's take the case of a Cu wire with a side length $x = 1$ mm. Assuming the highest temperature it can possibly operate T_{max} is the melting point of Cu (1357 K), we can work out the maximum current I_{max}. Note that we need to use the value of resistivity of Cu at this high temperature ($\rho \approx 10^{-7}\Omega$m) and we need to know a value for the emissivity of Cu which gives a k value of $\approx 2.5 \times 10^{-9}$ J s^{-1} m^{-2} K^{-4}.

$$I_{max} = 2\sqrt{\frac{kx^3}{\rho}}T_{max}^2 \approx 18\text{A} \tag{2.6}$$

current densities in superconductors to make much more compact and lighter magnets with fewer windings using less wire.

The other major selling point for superconducting magnets is the fact that they are much more energy efficient than conventional resistive electromagnets. To see why this is important, it is interesting to estimate how much power would be dissipated in a typical 3 tesla MRI magnet if it is made from 1×1 mm square copper wire operated at room temperature. If we assume that the magnet is a 1 m long ideal solenoid[3] and that the current flowing in the windings is 10 amps—safely below the maximum that it can withstand before it melts—then we would need about 240,000 turns of wire to generate the field. The internal diameter of the magnet needs to be big enough to be able to fit a person inside, so we are talking somewhere in the region of 700 km of wire, which will have a resistance of over 10,000 ohms. Power dissipated (P) depends on the current in the wire (I) and the resistance of the wire (R) according to the formula $P = I^2 R$, which comes out to be about a million watts (1 megawatt). To put that into perspective, one of those big wind turbines that you see on hillsides can typically generate about 2–3 megawatts of power and a household washing machine uses about 500 watts. By contrast, a superconducting magnet only requires power to keep it cold, and for a typical MRI magnet this turns out to be about the same as a few electric kettles—around 8000 watts (Herrmann and Rock, 2012).

2.2   How does magnetic resonance imaging work?

Magnetic resonance imaging (MRI) has been around since 1977 as an alternative to X-rays for imaging structures inside the human body. It has a lot going for it as a medical diagnostic technique. One of its advantages is that it is safer to use than X-rays. The reason that radiographers have to stand behind special lead-lined screens to take X-ray images is because high energy electromagnetic waves like these can ionise the atoms in our bodies and so exposure needs to be minimised. Another major benefit of MRI is that it can see soft tissue damage whereas X-rays can only see dense structures like bones. The main drawbacks are that the systems are more expensive and it takes considerably longer to take an MRI scan compared to a standard X-ray picture. MRI machines work in a completely different way to X-ray imaging. Instead of detecting the X-rays that manage to pass right through from one side of the body and out of the other side, they use magnetic fields to influence how the hydrogen atoms inside us behave. We are mainly made up of water—H_2O molecules—so we can get lots of useful information from what happens to the hydrogen atoms. The nucleus of a hydrogen atom is a single proton (positively charged particle) which has a special quantum mechanical property called *spin*. Although it is not really possible to visualise quantum mechanical phenomena, you can get some idea of what the spin of a particle is by imagining a spinning top. Protons (and electrons) do not actually spin like this

---

[3] An ideal solenoid is one that is infinitely long, so the magnetic field inside the bore is uniform and given by $B = \mu_0 n I$, where μ_0 is the permeability of free space ($4\pi \times 10^{-7}$ Hm^{-1}), n is the number of turns per unit length and I is the current. This is not really valid for the geometry of an MRI magnet, but it is good enough for a rough estimate.

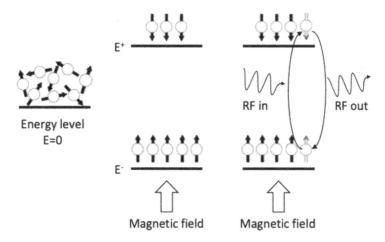

Fig. 2.3: Energy levels and alignment of spins with (a) no magnetic field applied, (b) with a static background field applied, and (c) when a radio frequency (RF) wave is switched on.

on their axis, but originally scientists thought that that was what was happening and the name stuck. This 'spin' makes each hydrogen nucleus act like a mini magnet.[4]

To understand how MRI works, we first need to think about how protons (hydrogen nuclei) behave in a magnetic field. If no magnetic field is applied, the magnetic poles of the hydrogen nuclei inside the material will point in random directions. When we apply a magnetic field, quantum mechanics tells us that the north pole of each proton will either point in the same direction as the magnetic field (parallel to the field) or will point in exactly the opposite direction (antiparallel to the field). There is some intrinsic uncertainty in knowing exactly how quantum mechanical particles behave, so we cannot actually know exactly what direction the north pole of each nucleus is pointing at any instant in time, but we know that *on average* it will be either parallel or antiparallel to the applied magnetic field. It turns out that the energy associated with the proton spin aligning parallel to the field (upwards in Fig. 2.3) is a little bit lower than aligning antiparallel to the field, so to save energy slightly more of the protons will have their spin pointing upwards than downwards. This means that the material will, on average, be magnetised in the upwards direction—parallel to the applied field. How strongly magnetic it becomes depends on how much field is applied because the energy difference between the up and down energy levels scales with the applied magnetic field. It is never a very strong effect though because out of every million protons, there are only around 10 more aligned spin up than spin down.

The basis of the MRI technique is to use an electromagnetic wave to persuade some of

---

[4]It is the spin of electrons that leads to atoms like Fe being magnetic. Electron spin is a much larger effect than nuclear spin, but in H_2O molecules the H atoms do not have any net electron spin, so it is only the nuclei that contribute to their magnetic moment.

the spin up protons to flip over and become spin down.[5] The electromagnetic wave is needed to provide the energy required to promote these protons from the lower energy level to the upper energy level—a process called *excitation*. It is really important that the wave has exactly the same energy as the energy gap between the lower and upper energy levels of the protons. If the energy is too high, or too low, the effect is very weak. This phenomenon of driving the system at exactly the right energy to produce the desired effect is what physicists call *resonance*, and we will come back to it in Chapter 9. In practice, for the kinds of magnetic fields used in MRI machines, to get resonance we need to use a radio frequency wave (RF) wave. This idea of exciting quantum mechanical particles from a low energy level to a high energy level may be familiar to you because it is the same thing that happens to electrons in atoms when they absorb light of certain frequencies and it can lead to objects having different colours. Importantly, the total energy of the excited system is higher than it was before, so there will be a tendency for protons in the upper level to flip back to lower the total energy, and as they do that they will emit the excess energy—again as a radio wave. This process is known as *relaxation*, and it is this radio wave that is detected in an MRI machine. If we keep the radio wave switched on, a dynamic situation develops where both excitation and relaxation processes occur, and a continuous radio wave will be emitted. The strength of the signal depends on how many hydrogen nuclei the material contains. However, if we apply a short RF pulse, the strength of the emitted radio wave will decay gradually, and the time it takes for the signal to drop off can be used to distinguish what kind of material the sample is made from. This is the basis of the NMR technique.

The quantum mechanical description gives us a good idea of the basis of the technique, but it can also be helpful to think about the classical picture of what is happening to the proton spins during the experiment. We can do this because the overall magnetic behaviour of the collection of protons will follow the rules of classical physics. We have already seen that a magnet will want to line up parallel with an applied magnetic field to lower its energy. You can do the experiment yourself using a bar magnet and a plotting compass (a small compass with a freely rotating magnetic needle). If you put the compass near the north pole of the magnet, the needle will point in the same direction as the magnet's north pole, but as you move the compass around, the needle will rotate to line up with the local direction of the magnetic field. The situation with an object that is rotating—spinning—is a bit different though. Instead of snapping into alignment with the magnetic field and just staying there stationary, its spin axis starts to rotate about the magnetic field direction and remains at an angle to it. This rotation is called *precession*, and it is exactly the same thing that happens in a gyroscope (Fig. 2.4).[6] The number of rotations in a second (the precession frequency)

[5] An electromagnetic wave consists of a fluctuating electric and magnetic field. There is a spectrum of electromagnetic waves with a wide range of different energies, with visible light sitting somewhere in the middle and radio waves having lower energy. The energy of the wave depends on its frequency—how many times the field fluctuates per second. In vacuum, all electromagnetic waves travel at the speed of light.

[6] A gyroscope is a special kind of spinning top which has a flywheel, and when you set it spinning, its spin axis starts rotating around the vertical direction of the gravitational field.

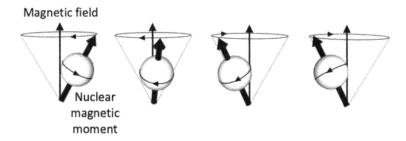

Fig. 2.4: Precession (rotation) of nuclear spins around the direction of the applied magnetic field.

is directly related to the strength of the applied magnetic field. So, if we set up a magnetic field that is completely uniform across a patient's body, the spin axes of all of their hydrogen nuclei will precess at exactly the same frequency about the direction of the applied field, and as they do so they will emit radio waves at that frequency.

So we have now got two different ways of thinking about the physics of magnetic resonance, but how do we use it to take images from specific regions of the body? The trick is to impose magnetic field gradients on top of the uniform background field. This makes it possible to pinpoint certain locations because only the protons in the correct magnetic field will be excited by our chosen frequency of radio wave. In practice, 3-D imaging is done using a series of stages, as illustrated schematically in Fig. 2.5. First, a small field gradient is applied along the direction of the main static magnetic field, so that the bottom of the apple in the figure is at a slightly lower magnetic field than the top. When the RF pulse is turned on, only the hydrogen nuclei in the slice that is exposed to the correct magnetic field to match the frequency of the radio wave will be excited and start to precess. But, if we just measured the intensity of this signal we would get an average from the whole slice (Fig. 2.5(b)). To get an image of the slice we need to apply field gradients in the other two directions after we have applied the RF pulse to start the spins precessing. If a field gradient is applied from left to right, the proton spins in the lower field region on the left hand side will rotate more slowly, and the ones in the higher field region on the right hand side will rotate more quickly. We can then use fairly standard mathematical processing methods to separate the detected signal into components of different frequency[7] and produce an image from each column of material within the slice. Separating the signal in the third direction requires another clever trick involving temporarily applying a field gradient in the third direction. We won't go into the details here because it just involves more complicated mathematical signal processing, but suffice it to say that it is not as easy to extract the information from the third dimension, and it requires a long sequence of separate measurements.

---

[7]We use something called a Fast Fourier Transform that converts a signal from the time domain to the frequency domain.

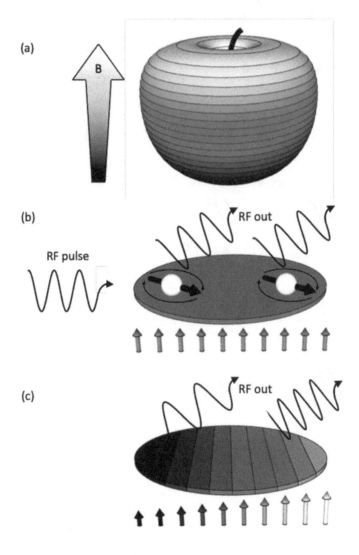

Fig. 2.5: Three-dimensional imaging in an MRI machine. (a) Initial application of the field gradient to allow a slice perpendicular to the field to be selected. (b) When no horizontal field gradient is applied, the signal generated by the precession of the spins in this slice is uniform. (c) Subsequent application of a field gradient from left to right affects the frequency of the signal emitted from different columns in the slice.

Superconductors are pretty much always used to generate the static background magnetic field in modern MRI machines. When you are wheeled into a scanner, you are actually lying inside the bore of a big superconducting solenoid operating at 4 K in liquid helium! MRI does not present much of a challenge for superconductors in terms of the raw field strengths required. Although state-of-the-art MRI for medical research can use magnetic fields of 7 tesla or higher, standard MRI machines found in hospitals typically use either 1.5 tesla or 3 tesla magnets. These field strengths could fairly easily be obtained with conventional copper magnets, although they would guzzle an awful lot of power. In fact, low resolution MRI machines can even use permanent magnets which are capable of generating fields of around 1 tesla. It is actually the *quality* of the magnetic field that can be achieved with superconductors that makes them essential for MRI. Here, by quality we mean both the uniformity and the stability of the field. In MRI it is super important to know exactly which point in space is in the resonant condition at any one point in time, so the background static field must be incredibly reliable. In conventional electromagnets, the problem with getting really stable magnetic fields is that the power supplies that drive the current fluctuate too much over time. Superconducting magnets have the huge advantage of being able to operate without a power supply at all, in what is called *persistent mode*. Because there is no resistance in a superconductor, once the current has been started a special superconducting switch can be closed to bypass the power supply and create a completely superconducting circuit. The power supply can then be removed altogether. It is no longer needed as the current will just carry on *ad infinitum*, provided there is no resistance anywhere in the circuit. This produces the kind of ultra-stable magnetic field that is perfect for MRI applications.

It is an added bonus that because there is no resistance in the circuit, no heat is generated, and because no heat is generated, evaporation of the liquid helium coolant is really slow. It is slow enough that a small mechanical cryocooler is all that is needed to recondense the helium as it boils off. This means that the energy required to cool the superconductor is a small fraction of the energy that would be needed to drive the current in a copper magnet, making them much more energy efficient and cheaper to run. In fact, MRI magnets are filled with liquid helium in the factory and shipped to the hospitals already cold. They very rarely need to be refilled in their entire 25 year lifetime, unless they need to be switched off and restarted for some reason. The ability to make magnets more compact by using superconductors is also an advantage in a hospital setting where space is at a premium.

2.3   Superconducting joints

In order for a superconducting magnet to operate in persistent mode without a power supply, we need the entire circuit to be superconducting. This means that we need at least one superconducting joint to close the circuit. In practice we need several more joints because a magnet uses a surprisingly long length of wire and it is not possible to manufacture single pieces long enough. Joining superconductors reliably without introducing any resistance is actually not as easy as it might sound. Part of the reason is that, as we have seen, the wires are not made from the superconductor alone.

Instead they are multifilamentary, consisting of many narrow threads of superconductor embedded in non-superconducting copper cladding (Fig. 1.5). This means that to make a fully superconducting joint, we first need to remove the copper to expose the superconducting filaments. Then we need to join them together somehow so that the superconducting current can make its way out of the filaments in one wire and into the filaments of the other.

The standard method for joining wires in the electronics industry is soldering. This works by temporarily melting a special filler metal called a solder so that it can flow and fill the gap between the wires. When the solder is left to solidify, an electrically conductive joint is formed. For obvious reasons, solder alloys need to have much lower melting points than the wires they are joining so that you do not melt the whole assembly during the soldering process! In the past, the most commonly used solder for the majority of applications was an alloy of lead and tin. This worked very well because both lead and tin are low melting point metals, and when you mix them together they make an even lower melting point alloy. This is rather similar to what happens when you add salt to water: the melting point of the water decreases, making it less likely to freeze. That is why we put salt on roads and paths in cold weather to stop them becoming icy. The other property that is important for solders is that they need to stick to the metal pieces that are being joined. The scientific name for this is *wettability* and lead-tin solders perform very well in this regard.

To make superconducting joints, we obviously have the extra constraint that the solder itself needs to superconduct at the 4 K operating temperature of the magnet because the current needs to pass through the solder to get between the wires. The joints are typically situated in locations outside the solenoid windings where the magnetic field is as low as possible (about 1 tesla for modern, compact MRI machines). Luckily for us, lead (Pb) is a pretty respectable superconductor—second only to niobium as far as the elements are concerned (see Table 2.1). Moreover, when we alloy it with other metallic elements we can improve its superconducting properties as well as lowering its melting point. Bismuth (Bi) turns out to be the favourite alloying addition for superconducting solders. Adding the bismuth pushes the critical temperature up a little, but most importantly it more than doubles the critical field so that it will still superconduct in the 1 tesla stray field from the magnet. There are several other alloys that also work pretty well, including standard lead-tin solder and Wood's metal which is a heady cocktail containing lead, bismuth, tin and cadmium.

There is a problem though. Lead is a highly toxic material, and its use in solders for most industries has now been outlawed. The consumer electronics industry, chemical industry and aerospace industry have developed new lead-free formulations which they now use exclusively. However, when it comes to looking for superconducting lead-free solder alternatives to replace lead-bismuth alloys, we run into a pretty serious problem. The superconducting properties of lead are far superior to any other low melting point element. As shown in Table 2.1, the next best choices are tin (Sn) and indium (In). Alloying these two together gives a material with a respectable critical temperature over 6 K, but the critical field remains well below the required 1 tesla. Adding in some bismuth to make a ternary alloy (one with three different elements), pushes

Table 2.1 Critical temperature (T_c) and critical field at zero K ($B_{c0}(0)$) values for selected elements and alloys.

Material	T_c (K)	$B_c(0)$ (T)
Nb	9.2	0.8
Pb	7.2	0.08
Hg	4	0.04
Sn	3.7	0.03
In	3.4	0.03
Pb-Bi	8.4	1.8
Sn-In	6.5	0.6
In-Bi	5.8	0.13
Sn-Sb	3.8	0.04
Sn-Bi	2.3	0.04

the performance up a little bit further, but nowhere near the heights offered by lead-bismuth solders. At the time of writing, the MRI industry is one of the few industries that are still legally allowed to use lead-based solders. Since the current exemption is due to expire soon, there is an urgent requirement to develop new kinds of reliable superconducting joints using solder-free techniques.

Aside from the environmental problem of needing to use a toxic lead-based alloy, making superconducting soldered joints ought to be relatively straightforward. After all, what could go wrong with such a commonly used and simple technique? The main problem turns out to be that when we remove the wire's copper cladding, by dunking it into an acid for example, the outside layer of the superconductor oxidises really quickly. This means that a layer of non-superconducting oxide coats the outside of each filament and acts as a barrier that superconducting current finds it difficult to get across. If the current has to pass through an oxide layer that is more than around 1 nanometre thick to get out of one wire and into another, it will put some resistance into the circuit. It turns out that for the NbTi superconductor that is used in MRI magnets, even 15 seconds exposure to air results in the formation of an oxide that is over 2 nanometres thick. Reliable, ultra-low resistance joints are so crucial for MRI that the practical tricks developed by magnet manufacturers to overcome these challenges are very closely guarded industrial secrets. If a single MRI superconducting joint fails, the entire magnet has to be shipped back to the factory, costing the company a vast amount of money as well as causing a great deal of inconvenience.

One common method that is really neat is called the *solder matrix replacement method*, illustrated schematically in Fig. 2.6. The word *matrix* is a materials science term for describing the main—usually interconnected—constituent of the material. In the case of multifilamentary wires, the copper cladding is the matrix because it surrounds the separated superconducting filaments. The boundaries (interfaces) between the copper cladding and the superconducting filaments tend to be nice and clean and free of oxide because of how the wire has been made (see Chapter 5). In the matrix replacement method, a few centimetres of the pristine wire are dunked into a bath of molten tin. The tin displaces the copper if you leave it for a few minutes, and when you remove

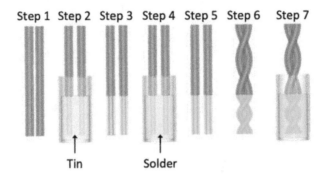

Fig. 2.6: The solder matrix replacement method used for making superconducting joints.

the wire, the filaments are surrounded by tin instead of copper. In the next step, the tin-coated wires are dunked into a bath of molten lead-bismuth solder. This displaces the tin and leaves superconducting filaments coated in superconducting solder. All of this happens without the surface of the filaments being exposed to air at all, so they do not get the chance to oxidise and the interfaces with the solder are clean. The solder-coated filaments from the two separate wires are carefully twisted together and usually pushed into a pot of solder which is left to solidify around the wires. You may wonder why we bother with the 'tinning' stage in the process at all since we then replace all of the tin with solder. The reason we do it is because tin is more effective than the solder at getting rid of the thick copper cladding. Tin is not a good superconductor though, which is why we need to use the two stage process.

As you can imagine, even using this clever method to minimise oxidation, it is unlikely that we can completely avoid introducing some resistance into the circuit at the joint. The question now is how much resistance can we get away with? MRI magnets need to have a field that is stable to less than 0.1 part per million per hour (Cosmus and Parizh, 2011). By this we mean that the magnetic field must change by a factor of less than 10^{-7} (0.00001%) over the course of an hour. For a standard 3 tesla magnet, this works out to be a tiny 0.3 microtesla in absolute terms. In the same way that friction introduces damping into a mechanical oscillator like a mass on a spring or a swinging pendulum, resistance in an electrical circuit will introduce damping. This leads to an exponential decay of the current (and therefore magnetic field) over time. The higher the resistance, the more damping and the more quickly the current decays. To achieve the stability required for MRI, it turns out that we typically need joint resistance values of less than 10^{-12} ohm (see Appendix B for the calculation). That sounds like a very small number, but let's try to put it into context. A 1 cm length copper wire with cross-sectional area of 1 mm^2 will have a resistance of about 10^{-4} ohm at room temperature—eight orders of magnitude higher than the resistance that we can afford to have in our joints!

2.4 What else can go wrong?

In addition to having to worry about resistance introduced into the magnet circuit by the joints, there are many other things that magnet design engineers have to take into consideration. For example, it is essential to make sure that the forces generated inside the magnet during operation are not too high. Otherwise the superconducting wire will break causing catastrophic failure. So where do these forces come from in the first place? You will already be very familiar with the idea that magnets exert forces on each other. If you put two magnets close together with the north pole of one magnet facing the south pole of the other, they will attract. If you turn one of the magnets around so the same poles are facing each other, then the magnets repel. You may also have come across the idea that if an object with an electrical charge moves in a magnetic field produced by a magnet of some sort, it will experience a *Lorentz force*. How big the Lorentz force is depends on how large the magnetic field is, how much charge the object has and how fast the charge is moving. The Lorentz force will always act in a direction that is at right angles to both the direction the charge is moving in and the direction of the magnetic field. So, if a positive charge travelling in the x-direction enters a magnetic field that points in the y-direction, it will experience a force in the z-direction that will change its trajectory. This is how the magnetic lenses in an electron microscope manage to focus an electron beam. It is also how a mass spectrometer manages to separate ions with different mass-to-charge ratio.

Since a current is, by definition, a flow of charge, it is not surprising that a wire carrying a current will also experience a Lorentz force if it is placed in a magnetic field. The force produced will be at right angles to both the current and the field. We can use *Fleming's left hand rule* to figure out which way the force will point by using the thumb of your left hand to be the direction of the force, your first finger to give the direction of the field, and your second finger to give the direction of the current (see Fig. 2.7). The equation for the Lorentz force is often written in the form $F = BIl$, where B is the magnetic field, I is the current and l is the length of the wire. I am pretty sure that I am not the only person who thinks of the Lorentz force as 'Bill'. In fact, this equation is only correct if the current is perpendicular to the field, because it turns out that the direction of the current relative to the direction of the magnetic

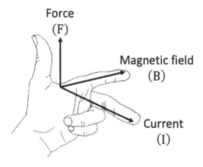

Fig. 2.7: Fleming's left hand rule.

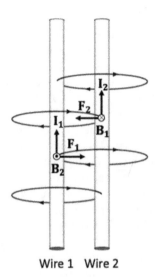

Fig. 2.8: Forces generated on two parallel wires carrying current.

field affects how big the resulting force is. The force is a maximum when the current and field are perpendicular, and the force will be zero if the current is parallel to the field.[8] Luckily, in the simple examples we will consider here, the field and the current are perpendicular to each other and so it is fine to use 'Bill' in its simple form.

Before we look at the Lorentz force in a solenoid coil like those used in MRI magnets, let's explore the simpler situation of two parallel wires carrying currents, as shown in Fig. 2.8. In 1819 Ampère discovered experimentally that if currents flow in the same direction in both wires they will be attracted to one another, whereas if currents flow in opposite directions the wires will repel each other.[9] The easiest way to analyse this situation is to consider how the magnetic field produced by one of the wires (wire 1) will be experienced by the other wire (wire 2). The field will circulate around the current in wire 1 in a right handed sense. By this I mean that if you point your right thumb along the direction of the current, the direction that your fingers curl will give the direction of the magnetic field. This means that the field produced by wire 1 will be pointing straight into the page at the location of wire 2. Then, using Fleming's left hand rule we can then see that the force on wire 2 (from the field generated by wire 1) will point left—towards wire 1. Using the same logic, the force acting on wire 1 due to the field produced by wire 2 will point towards wire 2. The force between the wires is attractive, consistent with Ampère's experimental findings.

---

[8] We can incorporate this angle dependence into the equation by including a factor of $\sin\theta$ where θ is the angle between the current and the field, or by writing it as a vector equation.

[9] It is this experiment that was used to define what 1 amp means. In modern units 1 amp is the current that needs to flow in infinitely long parallel wires 1 m apart to produce a force per unit length of wire of 2×10^{-7} Nm^{-1}.

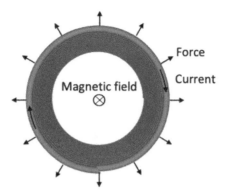

Fig. 2.9: Lorentz force acting outwards on solenoid windings.

So now we can return to the case of a solenoid magnet—a simple coil. We know that the magnetic field inside a solenoid will point straight along its axis. Specifically, if the current is circulating clockwise as we look down the axis of the solenoid, the magnetic field generated by the current will be pointing away from us (Fig. 2.9). This magnetic field will act on the current in the windings of the magnet itself. Using Fleming's left hand rule we can see that the force on any short section of the windings will always be acting radially outwards, away from the central axis. Therefore the solenoid will try to expand radially and a tensile (pulling) force will be generated in the wire. The size of the force scales with magnetic field strength, current and radius of the coil, and for a typical MRI magnet, the tensile force in the wire works out to be somewhere around 500 newtons. So are we in danger of breaking the wire? Tensile strength is the material's property that tells us how much *stress* (force per unit cross-sectional area) a particular material can withstand without breaking. It is of the order of 2 gigapascals for NbTi at 4 K (Koch and Easton, 1977). MRI wire has a cross-sectional area of about 1 mm^2 and it is subjected to a force of 500 newtons, which translates to a stress of around 0.5 gigapascals (5 $\times 10^8$ Pa)—not that far off the tensile strength of the wires. This means that magnet engineers need to include suitable strengthening structures in their designs, especially in the higher field magnets needed for particle accelerators or fusion reactors.

Under the Lens

Tension in windings

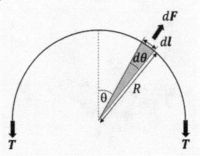

Fig. 2.10: Schematic for calculating tension in wire.

The Lorentz force (F) acting on a straight current (I) in a magnetic field (B) is often written as $F = BIl$ where l is the length of the wire carrying the current. This equation is only strictly true if I is perpendicular to B. Otherwise we have to take into account the angle θ between I and B.

$$F = BIl \sin \theta \qquad (2.7)$$

As you can see, when I and B are perpendicular, $\theta = 90°$ and the equation reduces to the earlier expression. In our magnet, we do not have a long straight wire though; we have a circular coil. To deal with this we imagine splitting the loop into small pieces, each of which is approximately straight. For an infinitesimally small length of wire dl (i.e. an infinitely short piece), we can re-write the element of force acting on this tiny length of wire as $dF = BIdl \sin \theta$. To calculate the tension in the wire, a neat way to do it is to imagine just half of a single loop of the wire as shown in Figure 2.10. To balance the tensile force T downwards, we need to add up all the upwards components of the Lorentz force. This is done by integrating over the correct angular range. In this case T will be balanced by the Lorentz force from a quarter of the loop, so we integrate from $\theta = 0$ to $\theta = \frac{\pi}{2}$ radians (90°).

$$dF = IBr \, d\theta \qquad (2.8)$$
$$dF \uparrow = IBr \, d\theta \cos \theta \qquad (2.9)$$
$$T = \int_0^{\frac{\pi}{2}} \cos \theta \, d\theta \qquad (2.10)$$
$$= IBr \left[\sin \theta \right]_0^{\frac{\pi}{2}} \qquad (2.11)$$
$$= IBr \qquad (2.12)$$

For a 3 T MRI magnet with a current of 500 A in the wire and a radius of 35 cm, this tensile force works out to be 525 N.

In addition to having to withstand some pretty large forces, MRI magnets typically store several million joules of energy in their magnetic field. This is equivalent to the kinetic energy of a double-decker bus travelling at about 50 mph. This is a potential safety hazard because if a little bit of the wire temporarily stops superconducting for some reason, some heat will be produced that might cause the wire close by to exceed the critical temperature as well. The situation would then avalanche and we would suddenly dump a huge resistance into the circuit. This is known as a *quench* and they can often be catastrophic because the energy stored in the magnetic field has to be dissipated very fast which can cause permanent damage to the magnet. They are pretty dramatic to watch because in a quench event the enormous current is deliberately diverted into an alternative (resistive) circuit where the energy is dissipated as safely as possible by boiling off all of the liquid helium in the storage tank. Helium explodes out of the pressure valves, forming a huge cloud of gas and making lots of noise! There are several things that can be done to try to stop quenches happening, including using clever electronics to pick up the early warning signs thus enabling evasive action to be taken. From a materials point of view, the reason that we put so much copper around the superconductor in the wires is to help minimise the chances of a quench. Copper is an excellent thermal conductor so it can carry heat away quickly in the early stages of a quench, reducing the chances of the hot-spot growing, and making it less likely that the entire wire cross-section will stop superconducting.

Interestingly, the fact that we can store a huge amount of energy in the magnetic field of a superconducting coil means we can use them as an efficient form of energy storage device. One of the problems with moving towards greener renewable energy sources is that they tend to be intermittent. Wind turbines will produce lots of energy when the wind conditions are right, but none at all when it is still, and if it is too windy they have to be switched off for safety reasons. Solar cells will generate much more energy on sunny days and will not produce any at all at night. This means that we need to have energy efficient ways of storing excess energy until we need it. There are lots of different options, from batteries to pumping water up to a high reservoir or storing gas under pressure in caves underground. Superconducting magnetic energy storage (SMES), where you store energy in a big superconducting coil, is one of the most efficient methods, achieving impressive cycle efficiencies above 95%. Coupled with that, it is also quick to recover the energy when it is needed and, unlike batteries, SMES systems can be charged and discharged an unlimited number of times. The drawbacks are the need for cryogenic cooling and the cost of the superconducting wire. Using high temperature superconductors reduces the cooling costs, but as we will see in Chapter 9, the conductor costs are much too high at present for these to be commercially viable. Several small SMES systems are used commercially for improving power quality and stability, and there are various larger scale test bed projects across the world that have demonstrated the feasibility of the technology.

Chapter summary

- Superconductors are the ultimate materials for making magnets because they can generate high magnetic fields with minimal energy expenditure in compact systems.

- MRI (and NMR) applications rely on obtaining very high magnetic field stability, which can only be achieved in superconducting magnets by operating in persistent mode without a power supply.

- One of the major technological challenges is making joints between wires that are fully superconducting to avoid introducing resistance into the circuit which would lead to unacceptable field decay.

- The tensile stresses generated in the magnets as a result of Lorentz forces can be pretty large, and this presents an engineering challenge, particularly as we move towards higher field magnets

- Superconducting magnets store a formidable amount of energy which can be exploited for energy storage applications, but also means that great care has to be taken to avoid catastrophic quenches.

3

Energy Essentials and Superconducting States

So far we have been thinking of superconductors as special materials that can carry electricity without losing any energy provided we keep them cold enough. We have seen that if we heat up a superconductor, it suddenly loses its zero resistance property and starts behaving as a normal metal with resistance. We also find that a superconductor stops working if we expose it to too large a magnetic field. If we exceed these limits, superconductivity will be lost, but we can recover it again by reducing the temperature and/or the magnetic field. This is because superconductivity is a special electromagnetic *state of matter*. The fact that we can switch between superconducting and non-superconducting states at will can be exploited in devices like very sensitive detectors of light or fancy fuses called fault current limiters that can reset themselves rather than burning out in the event of a fault.

Let's start by thinking about physical states of matter that you will already have come across—solids, liquids and gases. These different states of matter have different physical properties. For example, the liquid (molten) state of an ionic compound will conduct electricity, whereas the same material in the solid state will be an electrical insulator. Whether a substance is likely to be found in the solid, liquid or gas state at a specific combination of temperature and pressure depends on which of these states has the lowest energy.[1] Think about the familiar example of water at atmospheric pressure. Below 0°C it is most stable in the solid state as ice, but if we heat it up it will eventually undergo a transition into the liquid state—it will melt. This is because, above 0°C liquid water is more stable—has a lower energy—than ice. This fundamental idea that systems like to minimise their energy to become more stable will almost certainly not be new to you. For example, if you let go of a ball in mid-air it falls to the ground because the final state where the ball is at rest on the ground has a lower gravitational potential energy than the initial state when it was held at some distance above the ground. Similarly, if you let go of a stretched elastic band it springs back to its original length because this lowers its elastic potential energy. You may also have come across the idea that a chemical reaction will only occur spontaneously if the total energy of the products is lower than the total energy of the reactants.

---

[1]The kind of energy we are referring to here is the *free energy* of the substance which is formally a measure of the amount of work it can do at constant temperature.

In the same way that the liquid state of a material has different physical properties to the solid state of the same material, the superconducting state of a material has different properties to the normal state. Most obviously, the material only has zero resistivity in the superconducting state, but the magnetic properties and other physical properties like specific heat capacity[2] and thermal conductivity of the superconducting and normal states are also different. In this chapter we are going to look at the superconducting state and the superconducting-to-normal transition, with the purpose of discovering what fundamentally affects how good a superconducting material is. Along the way we will use *thermodynamics* and encounter *phase diagrams* which are vital tools for designing materials, as well as exploring *magnetic properties* of materials.

3.1   Thermodynamics

The study of the energy and its relationship to heat and work is called *thermodynamics*—a term that comes from the greek words 'therme' meaning 'heat' and 'dynamis' meaning 'power'. Despite having 'dynamic' in the name, it does not actually tell us anything about how quickly the system changes condition—that is dealt with in another branch of physics called *kinetics* that we will come back to in Chapter 8. Instead, it tells us what happens in a material when we change parameters like temperature and pressure slowly (*quasi-statically*[3]) by putting in or taking out energy in the form of heat or work. Let's consider a substance initially at some temperature and pressure. It will have a certain *internal energy* which is made up of the sum of its kinetic energy (from the movement of its atoms or molecules) and its potential energy (which arises from chemical bonding and other intermolecular forces in the material). If we change temperature and pressure to new values, its internal energy will change. It does not matter how the change is performed, only what the final state is. For example, if we first of all change the temperature and then subsequently change the pressure, the final state will have the same internal energy as if we had changed the pressure first followed by the temperature.[4]

Internal energy is an example of a thermodynamic *state function*, meaning that it gives us quantitative information about the state the system is in. Other state functions that you may have come across are *enthalpy*, *entropy* and *Gibb's free energy*. The values these properties take depend on what form the material is in (for example whether it is a solid, a liquid or a gas) as well as experimental conditions such as temperature, pressure and chemical composition. It is also worth noting that they are what are known as *extensive* thermodynamic properties because their values scale with the amount of the substance that is present—doubling the amount of the substance doubles the values of these state functions. For this reason we often define them per mole, per unit volume or per unit mass so that they become materials properties that are not affected by how much of the substance you have and so can be recorded in

---

[2]Specific heat capacity is the amount of heat needed to raise the temperature of 1 kilogram of the material by 1°C

[3]*quasi* is Latin for 'as if'. Changing variables quasi-statically means that we change them slowly enough that the system can respond in real time.

[4]This is the basis of Hess's heat cycles.

a data book! For example, to compare how heavy different materials are, we do not look up their mass because that depends on how big a chunk of the material we have. Instead we look up their density—mass per unit volume. If we double the volume of material, its mass will also double but its density is constant. Another extensive thermodynamic property that you may have come across is the heat capacity of a material which describes how much heat energy needs to be put into a material to raise its temperature by 1°C. The more material there is, the more energy is required, so data books give us the *specific heat capacity* which is defined as the heat capacity for one kilogram. In contrast, temperature and pressure are examples of *intensive variables* because they are ones that are not affected by the amount of substance that you have.

As materials scientists, we are often interested in what processes will naturally occur. We call these changes *spontaneous* and they happen when the final state is more stable than the initial state. But what exactly is the measure of stability? Earlier, I loosely related it to the energy of the system, but what kind of energy are we talking about? It would be tempting to think that a system with the lowest internal energy will be the most stable, but there are plenty of examples of spontaneous reactions in which the internal energy of the system increases. Another option could be that the system has to reduce its enthalpy. The enthalpy change of a reaction is essentially a measure of the heat change at constant pressure, and so a reaction accompanied by a reduction in enthalpy gives out heat to the surroundings—it is exothermic. If you have done some chemistry, you will probably be aware that indeed a lot of spontaneous reactions are exothermic, but some are endothermic—they take in heat from the surroundings. So enthalpy does not give us a fail safe method for predicting spontaneous changes.

In fact, the second law of thermodynamics tells us the condition for spontaneous change: the *entropy of the universe* must increase. Here, the 'universe' relates to the system (the material or reaction we are considering) together with its surroundings. Entropy is a measure of disorder. The more disordered something is, the higher its entropy will be. The idea that more disordered things are more stable is not immediately obvious, but you can get some idea by thinking how easy it is for your house to spontaneously get messy, and how much effort you have to put in to tidy it again! Gases are naturally more disordered than liquids and solids, so you would be forgiven for thinking that solids will never be stable. Luckily this is not the case because a *system* can spontaneously become more ordered, provided the *surroundings* make up for it by becoming more disordered. However, it is not very convenient to have to consider what is happening in the whole universe, so we use a different state function called the *Gibb's free energy* of the system, which will always decrease in a spontaneous reaction. The most stable state for a system under certain conditions will be the state that has the lowest Gibb's free energy. If two states are in equilibrium—for example liquid and steam at the boiling point of water—they will have the same Gibb's free energy.

Under the Lens

Gibb's free energy

The second law of thermodynamics tells us that spontaneous reactions are accompanied by an increase in the entropy of the universe. Let's break that down by looking at the solidification process where the system is a liquid that is being converted to a solid in some open container, and the surroundings are the atmosphere. Solidification is an exothermic process—it gives out latent heat (enthalpy change associated with solidification). That heat is transferred to the surroundings, making the gas molecules in the air move more quickly and increasing their disorder. In fact, under constant pressure conditions the entropy change of the surroundings ($\Delta S_{surroundings}$) is related to the enthalpy change of the system (ΔH_{system}) by the simple formula $\Delta S_{surroundings} = -T\Delta H_{system}$, where T is temperature. This means that although solidification decreases the entropy of the system (the solid is more ordered than the liquid), the entropy of the surroundings increases as a result of the heat released during solidification. If this increase in entropy of the surroundings is larger than the decrease in entropy of the system, then the universe has increased its entropy overall and the process will happen spontaneously.

By relating the entropy increase of the surroundings to the enthalpy change of the system (how much heat is being given out), it is possible to define a thermodynamic state function—the Gibbs free energy ($G = H - TS$) that tells us whether a constant pressure process will be spontaneous or not. Starting with the second law of thermodynamics, the condition for a spontaneous change to occur is given by

$$\Delta S_{universe} > 0$$
$$\Delta S_{surroundings} + \Delta S_{system} > 0$$
$$-T\Delta H_{system} + \Delta S_{system} > 0$$
$$\Delta H_{system} - T\Delta S_{system} < 0$$
$$\Delta G < 0 \tag{3.1}$$

This result shows that any spontaneous process (under constant pressure conditions) will be accompanied by a decrease in Gibbs free energy. A different kind of free energy—the Helmholtz free energy—is defined in a similar way for processes that occur under constant volume conditions. Materials scientists usually work with Gibbs free energy because many reactions that we are interested in occur at room pressure.

3.2   Phase diagrams

Materials scientists often draw special diagrams called *phase diagrams* to show the thermodynamically stable form (or forms) of a system under different conditions. Looking at the simple phase diagram for water (Fig. 3.1), you would be forgiven for thinking that a *phase* is exactly equivalent to a *state of matter* because the three forms of water shown are solid, liquid and gas. In fact, there are subtle differences in their definitions. Specifically, a phase is a form of matter that has uniform chemical and physical properties, which means that different states of matter are necessarily different phases. Interestingly, the statement is not true the other way around: different phases are not always in different states, but we will come back to this in Chapter 5.

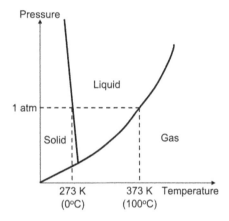

Fig. 3.1: Phase diagram of water.

Returning to the relatively simple case of water. It can exist in three different phases—ice, water and steam—each corresponding to a different state. Here, the only two variables of interest are temperature and pressure, but in general there are other variables that may be of importance. For example, you will see in Chapter 5 that chemical composition is important in alloy systems like Nb-Ti. For superconductors and other magnetic materials, the strength of the applied magnetic field is also a variable that must be considered. Of course, the more variables we have to consider, the more complicated the diagrams become because we need a different axis for each parameter that we are changing! We can use phase diagrams to predict what will happen when we change the conditions. Starting at low temperature and atmospheric pressure (1 atm), the phase diagram tells us that the solid phase (ice) will be the most stable. We can melt the ice by increasing the temperature (moving right on the diagram). We can also melt the ice by increasing pressure (moving upwards on the diagram) because water is unusual in that it is denser in the liquid state than in the solid state. For most other materials, increasing the pressure would stabilise the solid state. At conditions along the boundary between the liquid and solid phases, both phases will happily coexist with each other. Out of interest, there is one particularly special point on the phase diagram of water—the point at which all three phase regions meet each other. This is called the *triple point* (for fairly obvious reasons), and at this specific temperature and pressure the solid, liquid and gas phases are expected to all coexist together.

It is worth mentioning that phase diagrams only tell us what we would expect to happen at *equilibrium*—in the thermodynamically most stable condition.[5] Whether or not the material will be in its equilibrium condition will depend on how easily the system can change from one phase to another in practice and its thermal history. For example, it is quite common for a liquid that is cooled quickly not to start solidifying until it reaches a temperature considerably below its equilibrium melting point.

3.3   The superconducting transition

Now, we are going to apply these concepts to explore the electromagnetic states in superconducting materials. Figure 3.2 shows the magnetic phase diagram of a typical superconducting material of a fixed chemical composition at fixed pressure. As you can see, providing the combination of temperature and magnetic field are kept below the curved line, the material will be in the superconducting state. However, above the line the energetic advantage offered by superconductivity is lost and the material prefers to be a normal resistive metal. We call this state the *normal state* of a superconductor. We have not changed its chemical constitution, and the material is still a solid, but we have changed its thermodynamic state and its electromagnetic properties.

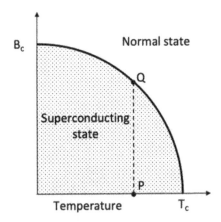

Fig. 3.2: Magnetic phase diagram of a generic superconductor.

Let's explore the magnetic phase diagram in more detail. If we start at the point on the diagram labelled P (with zero magnetic field applied) and increase the temperature (move to the right), the material stays in the superconducting state until it reaches its critical temperature, T_c. At this temperature superconductivity would immediately

---

[5] At equilibrium, the Gibb's free energy of the system is minimised.

be lost as soon as you increase the magnetic field by even the tiniest amount (move upwards on the diagram). However, if we go back to the lower temperature at point P, we can now increase the magnetic field up to point Q before losing superconductivity. The strength of the magnetic field that can be tolerated is called the critical field (B_c) and its value gradually decreases as temperature is increased, until at T_c, B_c has dropped to zero. There is a subtle difference between how we refer to the critical field and the critical temperature. Although T_c always means the critical temperature in zero magnetic field, we have to specify the measurement temperature when we quote critical field values. Often we state the values at 4.2 K—the boiling point of helium—because this is the most common operating temperature. The other value that physicists like to quote to compare one material with another is what the critical field would be if we could reach absolute zero (0 K). Of course we cannot actually do this in practice, but it turns out to be a useful way to compare different materials and it is used in many theoretical treatments.

3.4 Magnetic properties of superconductors

Before we go on to consider the thermodynamics of the superconducting-to-normal phase transition in more detail, it is interesting to explore what actually happens when we put a superconductor into a magnetic field. As well as having the extraordinary property of zero resistance, superconductors have unique magnetic properties that are responsible for whacky phenomena like magnetic levitation. But what is so special about the way a superconductor responds in a magnetic field? Unlike any other material, the simplest kinds of superconductors are able to prevent magnetic field getting into their interior. Imagine a sphere of superconductor inside the bore of an electromagnet in which we can control the temperature. Let's start with the magnet switched off and at a temperature above T_c so the material is in its non-superconducting normal state (as shown in diagram A in Fig. 3.3). If we turn on the electromagnet, we generate a magnetic field that passes straight through the material because in the normal state it is essentially non-magnetic and will not interact with the magnetic field at all (B). Now, a remarkable thing happens when we cool the system into the superconducting state. The magnetic field that was penetrating the material gets completely pushed out (expelled) from the superconductor, leaving it with essentially zero field inside (C).[6] This is known as the *Meissner effect*. Interestingly, we would reach the same final state with the magnetic field fully excluded if we first cool the material in zero field (D) and then switch on the magnetic field. This is a necessary requirement of a thermodynamic state—its properties under certain conditions depend only on what those conditions are, not the path which we have taken to get there.

So, how does a superconductor expel the magnetic field? In fact, it cleverly creates its own magnetic field in the opposite sense to the field applied by the external magnet, meaning that the field is completely cancelled out inside the material. Again, this begs the follow-up question—how? Remember in Section 2.1 we talked about currents circulating in solenoids (coils of wire) producing uniform magnetic fields within their bore. Well, if we take a cylindrical piece of superconductor, a similar effect would be

[6]Except in a narrow shell at the surface.

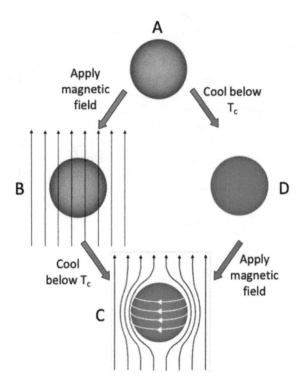

Fig. 3.3: Behaviour of superconductors when magnetic field is applied before and after cooling into the superconducting state.

produced if currents were made to flow in circular paths around its curved surface (see Fig. 3.4). This is exactly what happens when we put a superconductor in an external magnetic field. If the applied magnetic field points vertically upwards along the axis of the cylindrical sample, currents would automatically start to flow around the outside surface of the superconducting cylinder in the correct sense so as to produce the same magnitude of magnetic field vertically downwards. So does this only happen in superconductors? In fact, any electrical conductor that is initially in zero magnetic field will subsequently attempt to stop any applied field from entering. This is a manifestation of Lenz's law that you might have come across when learning about Faraday's law of electromagnetic induction. While the external magnetic field is being changed, an electromotive force is generated that will set up circulating currents in the surface of the conductor to try to stop the field inside it changing.[7] The problem is, because normal conductors have resistance, these eddy currents decay (drop off) very quickly when the external field stops changing and the electromotive force disappears. This means that the magnetic field almost immediately enters the material. It is the zero resistance property of a superconductor that means that the currents that start

---

[7]This effect is responsible for the fact that if you drop a permanent magnet through a copper tube it will take much longer than you expect to come out of the bottom.

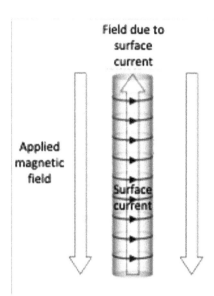

Fig. 3.4: Circulating surface currents screening the applied magnetic field.

flowing when the field is changing persist even when the field stops changing and the electromotive force becomes zero again.[8]

In most materials that we think of as being magnetic, like iron, the material produces a magnetic field that is in the same direction as the externally applied field, rather than the opposite direction. This is because the atoms inside the material behave as mini magnets which we refer to as *atomic magnetic dipoles* because they each have their own north and a south pole. We typically represent them schematically as little arrows pointing north. In the simplest kind of magnetic material, the atomic dipoles are aligned in random directions until the material is placed in an external magnetic field, at which point the dipoles like to rotate to line up with the applied field. When the field is removed, the dipoles become randomly aligned again. This effect is known as *paramagnetism* and it usually only produces a weak magnetic response. Iron, which has a stronger magnetic response, is a different type of magnetic material called a *ferromagnet*. In ferromagnets, the atomic dipoles are strong enough that they choose to line up in the same direction as their neighbours, even when no external magnetic field is present. It is this spontaneous alignment of atomic dipoles that is exploited in permanent magnets like the ones we use to stick notes to the fridge. Interestingly though, chunks of iron and other ferromagnetic materials are not naturally magnetised because there is an energy cost associated with generating the large magnetic field in the region outside the magnet. Instead, there is a tendency for a ferromagnet to split into small regions called *magnetic domains*. Within a single domain the atomic dipoles are perfectly aligned with each other, but adjacent domains are typically aligned in

---

[8]This is exactly the same phenomenon that we exploit in MRI magnets that operate in persistent mode without a power supply.

opposite directions, meaning that the material will not be magnetic overall. When the material is placed in a magnetic field, there is an energy saving by having the atomic moments lined up with it and so domains that have their magnetic moments pointing in the same direction as the applied field grow, and unfavourably aligned domains shrink. In a pure piece of iron, which is a soft ferromagnet, the magnetisation is lost when the external field is removed because the material reverts back to something like the original domain structure. However, in a hard ferromagnet like steel, it is quite difficult for the domains to move back again, and so the material stays magnetised to some extent even when the external field is removed. It is these magnetised hard ferromagnets that we usually refer to as *magnets*. In general usage, we say that any material that is attracted to a magnet is *magnetic*, but scientifically any material that responds in any way to a magnetic field is magnetic—including a superconductor.

It is useful at this point to define some terms that tell us about how strong the magnetic response of a material is. The most straightforward measure is the total (net) magnetic moment of a sample, which we find by simply adding up all of the atomic magnetic moment pointing in one direction and subtracting all of the atomic moments pointing in the opposite direction. However, this quantity will depend on how much material we have in total—how many atomic dipoles there are in our sample. It turns out to be more convenient to use a quantity that is independent of sample size, so we introduce the term *magnetisation*, which is formally defined as the total magnetic dipole moment of the sample divided by its volume. In paramagnetic or ferromagnetic materials, we would expect magnetisation to increase as we increase the external magnetic field until all of the dipoles are aligned perfectly. This means that to compare how easy it is to magnetise different materials, we need to define another property that tells us how much the magnetisation of the material changes for a given change in magnetic field. This materials property is called the *magnetic susceptibility* and for both paramagnets and ferromagnets it will have a positive sign because they become magnetised in the same direction as the applied magnetic field.

In a superconductor, the situation is completely different because their magnetic response is not due to permanent magnetic dipoles on the individual atoms. Instead of the material becoming magnetised in the same direction as the applied magnetic field, surface currents circulate around a superconductor producing a magnetic field that opposes the applied field. This leads to a magnetisation pointing in the opposite direction to the applied magnetic field. The general term used to describe materials that show this kind of negative susceptibility is *diamagnetism*, and superconductors are known as perfect diamagnets because (under certain circumstances) sufficient current can flow to perfectly cancel out the applied field. Superconductors can expel all of the applied field, whereas in other diamagnetic materials the effect is very weak and only a very small fraction of the applied field is expelled.

3.5 Why is there a critical field?

We have already come across the idea that superconductors only work up to some maximum magnetic field called the critical field, B_c. If we place the material into a magnetic field that is too high, it stops superconducting and becomes normal. So why does this happen? We know that at temperatures below the critical temperature the superconducting state is stable and so it must have a lower energy than the normal state. The energy saved (when no magnetic field is applied) is called the *condensation energy*. However, when we apply a magnetic field to a superconductor the material has to do some work to push the magnetic field out of the material. For those of you that know some thermodynamics already, this is the magnetic equivalent of the compression stage in a heat engine cycle when the temperature is kept constant but mechanical work is done against pressure to squash the gas into a smaller volume. Not surprisingly, the larger the magnetic field, the more energy has to be put in to expel it. Therefore, as we increase the applied magnetic field the energy cost increases and eventually it will balance out the condensation energy that has been saved in the first place by it being in the superconducting state. If we continue to increase the applied field, the magnetic energy cost would outweigh the condensation energy so the material would save itself energy by reverting to the normal state and letting all the magnetic field flood back in. (Remember, in the normal state, magnetic field is not excluded so we do not have to do any magnetic work anymore). The critical field is therefore the field at which all of the condensation energy saved by the material being in the superconducting state is used to do magnetic work to expel the magnetic field. It turns out that the condensation energy of a superconductor scales with the square of its critical field (see Appendix C).

You may have noticed that we have not, at this point, attempted to explain the origin of the condensation energy—why the superconducting state has a lower energy than the normal state at temperatures below T_c. This is because it requires us to think about what is happening inside a superconductor at the microscopic scale. In particular we need to think about what happens to the electrons in the material, and we are leaving that discussion to Chapter 6.

The Wider View

Phase transformations in materials science

The thermodynamics approach that we have used to understand the superconducting transition can be equally well applied to other phase changes. This is very handy for materials scientists because we often want to understand what happens when one solid phase converts to another solid phase inside a material or what happens when we use melting or vaporisation in the manufacture of artefacts. One obvious example is the casting of solid chunks of metal by melting them and pouring them into moulds. This is used at some point in the manufacturing process of the vast majority of metallic (and plastic) objects.

A slightly less familiar example is the growth of thin films—coatings. Thermal evaporation is a commonly used method for depositing thin layers of, say, aluminium (Al) on another material, and is shown schematically in Figure 3.5. This process works by putting a solid block of Al metal (the source) into a special ceramic cup called a crucible and placing it at the bottom of a vacuum chamber. A 'substrate'—the thing that we want to grow the film onto—is placed at the top of the chamber. The air is pumped out of the chamber until the pressure is much lower than atmospheric pressure and then the crucible is heated up causing the Al to evaporate. The vaporised Al atoms drift towards the substrate and condense to form a solid film. By changing the temperature of the Al source, it is possible to control the pressure of the Al vapour and therefore how quickly the deposition takes place. Interestingly, aluminium is a (not very good) superconductor, and it is frequently used to make superconducting devices simply because it is so easy to make reliably as thin film coatings, as we will come back to in Chapter 6.

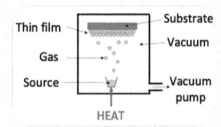

Fig. 3.5: Thermal evaporation process.

3.6 Heat capacity at the superconducting transition

The heat capacity of a material—how much heat energy it takes to change its temperature by 1 K—is a much easier thermodynamic property to measure than state functions like Gibb's free energy or internal energy. Therefore it is interesting to explore what we would expect to happen to the heat capacity when we cool a material through the superconducting transition. It turns out that the heat capacity of the superconducting state is higher than that of the normal state, so there will be a sudden increase in heat capacity when the material turns superconducting, as shown in Fig. 3.6.[9] This means that heat capacity measurements are an excellent way to spot a superconducting transition and measure T_c. In fact, it turns out that this jump in heat capacity is such a good signature of superconductivity that it is one of the three things that you need to prove if you discover a new superconductor—the other two being zero resistivity and perfect diamagnetism.

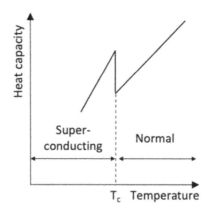

Fig. 3.6: Jump in heat capacity at the superconducting transition temperature.

There is an interesting related story about the surprising discovery of superconductivity in magnesium diboride in 2001 that we will come back to in Chapter 10. The real surprise was why nobody had discovered it before. Not only is it one of the simplest superconductors chemically, containing just two elements, but it has been manufactured in large quantities since the early 1950s and it turns out that it has a relatively high transition temperature of 39 K. The reason it was not discovered earlier is really because nobody expected that it would superconduct so nobody looked for it. In fact, data published as early as 1957, in tabular form, shows an anomaly in heat capacity at around T_c (Swift and White, 1957), so the clue was there 44 years before superconductivity in MgB_2 was reported!

---

[9]Strictly speaking, this is only true in low magnetic fields.

3.7   Type I and type II superconductivity

So far, we have talked about the most simple type of superconducting state, where all of the magnetic field is expelled from the inside of the material. We often refer to this special state as the 'Meissner state' of a superconductor. For some materials—usually simple elemental metals like pure lead—if we ignore some wacky effects that can occur because of the shape of the sample,[10] the superconductor will be in the Meissner state for all magnetic fields up to the critical field. Above the critical field, superconductivity is destroyed everywhere and the entire material enters the normal state. These materials are referred to as *type I* superconductors. They are not very useful because they have low critical temperatures, often below the boiling point of liquid helium, as well as having very low critical fields—usually low enough that superconductivity can be destroyed with a standard fridge magnet. Luckily, there is more to the story, otherwise superconductivity would have been consigned to the relics of history as nothing more than an interesting physical phenomenon with no practical use.

Not long after the original discovery, it was realised that superconductivity was not reserved just for pure elements. It could also happen in alloys—mixtures of two or more metallic elements—and this time the materials seemed to remain superconducting in much higher magnetic fields. For instance, lead is one of the best elemental superconductors with a critical field of about 0.08 tesla and elemental bismuth is a very poor superconductor. When the two are alloyed together to form $Pb_{60}Bi_{40}$, a common superconducting solder, the critical field rises to about 1.8 tesla—more than twenty times higher than pure lead. So why can the metal alloys do so much better than the pure metallic elements? Either they must have much larger condensation energies for some reason (so that they can thermodynamically afford to push out more magnetic field) or something else must be going on. It turns out that the latter is true, and we refer to this new class of materials as *type II* (type two) superconductors.[11]

Let's explore the differences between these two types of superconductor by comparing what happens when we put a type I and a type II superconductor in a magnetic field. For simplicity, we are going to assume our piece of superconductor is a cylindrical rod and we are applying the magnetic field along its axis. This means that we do not have to worry about any field concentration effects caused by the shape of the sample. Figure 3.7 shows what we know already about a type I superconductor. At magnetic fields below the critical field (B_c), the field is completely expelled from the superconductor so the internal field is zero. When we go above B_c superconductivity is lost altogether and the external magnetic field is immediately allowed back into the material. We can also think about what happens to the magnetisation of the material as we apply the external field. In the superconducting state the superconductor has a negative magnetisation because the material is producing a magnetic field that is equal

---

[10]The concentration of magnetic field outside the superconductor for some sample shapes can result in the formation of the *intermediate state* in which some regions of the material turn normal before superconductivity is lost everywhere.

[11]It is worth noting that commercial niobium alloys NbTi and Nb_3Sn also have much higher critical field values than pure niobium, but niobium itself also has an unusually high critical field for elements (0.8 tesla). This is because even pure niobium is a type II superconductor.

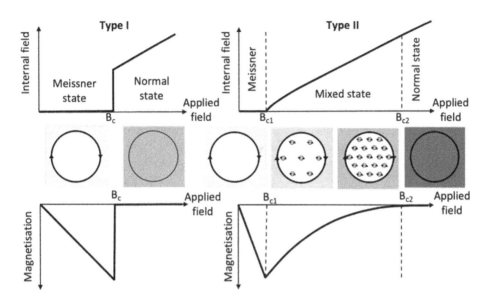

Fig. 3.7: Comparison between magnetic behaviour of type I and type II superconductors. The magnetic field distribution is shown schematically in the diagrams, with
darker grey indicating a higher field. The circulating currents are also indicated.

and opposite to the externally applied magnetic field. As the applied field increases, its
magnetisation will gradually become more negative. When the material turns normal,
the magnetisation is suddenly lost because it is no longer energetically favourable to
expel the field.

Now let's compare that with the magnetic behaviour of type II superconductors. To
begin with, at low applied fields, the material responds in exactly the same way as
a type I superconductor: the field is completely expelled and the magnetisation gets
steadily more negative as the field is increased. The superconductor is in the Meissner
state. However, there comes a point called the *lower critical field* (B_{c1}) where the material stops excluding all the magnetic field and starts to let a little bit of it get inside.
Instead of letting all the field flood in at once, as happens in type I superconductors,
it enters little by little. As it does so, the material becomes gradually less strongly
magnetised (its magnetisation gets less negative) because it is not excluding so much
of the applied magnetic field. It is not until it reaches a much higher field—the *upper
critical field* (B_{c2})—that the material loses superconductivity completely.

So why does this happen? Well, if we think about it, it is quite a sensible thing to do
from an energetics point of view. By allowing some magnetic field inside, the superconductor can save some of the energy cost associated with keeping it out. The problem is
that we still are not allowed to have magnetic field inside the superconducting phase.
What happens instead is that narrow tubes are created in the superconductor that
act as channels that the magnetic field can pass through. They are not actually physical holes, but are actually long, thin cylinders of normal (non-superconducting) phase

that form spontaneously within the material. We have various names for these features including 'flux lines', 'flux tubes' or 'fluxons', where the term *flux* refers to the amount of magnetic field passing through each one.[12] They are sometimes even called 'flux vortices' because the magnetic flux passing through their core is produced by currents circulating like a vortex or eddy. In fact, flux lines are rather like tornadoes in structure—the wind circulating around the eye of a tornado is like the current circulating around the normal core. There is an energy cost associated with making flux lines though, because we have to overcome the condensation energy to transform the volume of material that makes up the core of a flux line from the superconducting to the normal state. So, on the one hand the system can save magnetic energy by letting some field get inside the material, but on the other hand it costs condensation energy to produce the normal tunnels through which the magnetic field can pass. Whether or not we end up with a type I or a type II superconductor essentially comes down to which energy term wins. The details are rather complicated because the walls of the 'tunnels' are not rigid so the volume of normal material that is created is not the same as the volume through which the magnetic field is allowed to pass.

Without getting too bogged down in the details of the fundamental physics, it turns out that there is an *interface energy* associated with boundaries between normal and superconducting phases (the walls of the flux tubes). This is an energy term that scales with the area of the boundary, or interface, between the two phases. The idea that interfaces have an energy associated with them is pretty much universal. Liquid/solid interfaces have an energy. Boundaries between two different crystals have an energy and interfaces between two different solid phases also have an energy. Moreover, all of these interfaces have *positive* energies associated with them—it costs us energy to create the interface. This is because the interface is a region of the material where the bonding and arrangement of the atoms is not perfect. It is for this very reason that it turns out to be difficult to engineer materials with really small scale features—there is a driving force for them to grow (coarsen) because larger particles have a smaller surface area to volume ratio, thereby saving interface energy.

In type I superconductors the interface energy between normal and superconducting regions is also *positive*—it costs energy to create the boundary—and so flux lines are not stable. By contrast, the normal/superconductor interface energy in type II superconductors is rather unique because it turns out to be *negative*. The system actually saves energy by creating these interfaces. Unusually, the negative interface energy drives the system to form as large a surface area as possible between normal and superconducting regions. Instead of all of the magnetic flux penetrating through a single tube, it would be better to split the flux into two tubes because this would result in a larger interface area. Better still, split it again and have four tubes, each with a quarter of the total flux. Repeating this splitting process will continue to save interface energy, and so each tube actually contains a single *magnetic flux quantum* (Φ_0). This is the minimum packet of magnetic flux allowed by quantum mechanics and

---

[12]Remember that the magnetic field (B) is often referred to as a 'magnetic flux density' and is essentially the amount of magnetic flux passing through an area of 1 m^2.

it has a value of $\Phi_0 = 2.07 \times 10^{-15}$ Wb. We will come back to the idea that physical properties comes in discrete packets—quanta—in Chapter 6.

Between the lower critical field (B_{c1}) and the upper critical field (B_{c2}), the superconductor is said to be in the *mixed state*—a two phase system consisting of normal and superconducting regions coexisting with each other. As we increase the applied magnetic field above B_{c1} the magnetic flux density inside the material is seen to gradually increase. To achieve this the *number* of flux lines increases (rather than how much flux goes through each one) to maximise interface area and therefore minimise the total energy of the system. The flux lines get closer and closer together and the volume fraction of normal material gradually increases until, at the upper critical field (B_{c2}), the normal cores start to overlap with each other and the whole material becomes normal. This is shown schematically in Fig. 3.7. An important point to appreciate is that B_{c2} values are typically several orders of magnitude higher than B_{c1} values. For example, in Nb$_3$Sn the lower critical field is only 0.04 tesla, which is about 0.2 % of its upper critical field. Since B_{c1} values are so low, in almost all practical applications of superconductors, the materials operate in the mixed state (not the Meissner state). As we will see in Chapter 4, this has major implications for the definition of critical current in type II superconductors and allows us to use all sorts of materials science tricks to optimise the current carrying capacity.

Chapter summary

- Superconductivity comes about because, under certain conditions, it is the lowest energy (most stable) state of the material. In order to understand what is going on energetically inside the materials we have used the laws of thermodynamics.

- Phase diagrams are a vital tool for materials scientists because they help us to work out how to design materials and processing strategies to get the optimum internal structure and therefore the best materials performance. We will encounter several more examples of this later in the book.

- The idea that superconductors come in two flavours, type I and type II, has been introduced from the perspective of there being competing energy terms within the material.

- With the possible exception of some small scale device materials, all practical superconductors are type II materials and under all useful operating conditions will therefore have non-superconducting flux lines running through them. We can control the behaviour of these flux lines by engineering the materials very carefully on the nanoscale. This is the crux of why the materials science of superconductors is so interesting and important.

4

Levitation Magic

Have you ever wished that you could fly on a magic carpet or ride a hoverboard like Marty McFly in *Back to the Future II*? Maybe you would like to be able to make objects gracefully rise from the table like Harry Potter and his friends, by uttering the famous words 'Wingardium Leviosa'. There is something utterly magical and captivating about the idea of making objects defy the laws of gravity and float in mid-air. But this is just the stuff of science fiction. The super-cool hoverboard that featured in *Back to the Future II*— in 1989 was indeed just a special effect, although the film's director, Robert Zemeckis, did fool the public for a while by pretending that it was real! Since then there have been several high profile hoverboard hoaxes using computer trickery to give the illusion of levitation. But despite all the hype, the idea was not as ridiculous and 'futuristic' as you may think. As it happens, high temperature superconductors had just been discovered when the film was released, and these materials turn out to be the key to making practical levitation a reality.

The basic physics behind levitation is actually quite familiar to us. If you have ever played with fridge magnets, you will probably know that when you bring two magnets together, what happens depends on which way round the magnets are. If you bring the north pole of one magnet close to the south pole of the other magnet, they will stick together. However, if you turn one of the magnets over, so you bring them towards each other with their north poles facing, they will repel. Magnetic levitation, in its simplest form, exploits this effect—you do not even need particularly strong magnets for the repulsive force to be large enough to support the magnet's own weight. There is one major problem though. If we have one fixed magnet and try to float a second magnet above it, the top magnet will immediately flip itself over and stick fast to bottom one. The only way to stop this happening is to somehow constrain the 'floating' magnet so that it cannot flip over. This fidgety behaviour makes it pretty difficult to harness the levitation power, seriously limiting how useful the technology is. You would not want to be on a hoverboard that is perpetually trying to kick you off!

Superconductors turn out to be a real game-changer for practical levitation because of the unique way they respond to being placed in a magnetic field. Just like many other magnetic materials, a superconductor does not generally behave as a permanent magnet. Instead, it only becomes magnetised when it is placed in a magnetic field created by something else. What makes superconductors really special is that, unlike most familiar magnetic materials like pure iron, they are diamagnetic. They always becomes magnetised in the *opposite* direction to the magnet. So if the magnet is placed on a table with its north pole facing upwards, the superconductor becomes magnetised with its north pole pointing downwards. This means that a superconductor is naturally repelled from a magnet rather than being attracted to it. If we flip the superconductor over, it will still be repelled because the material actively responds to make sure its magnetisation always opposes that of the magnet. If you flipped the magnet over so its south pole faced upwards, the superconductor would instantaneously respond by magnetising itself in such a way as to have its south pole facing downwards. It does this because, in their Meissner state at least, superconductors expel all of the magnetic field and stop it getting inside the material. As discussed in Section 3.4, they magnetise themselves in such a way so as to produce a magnetic field that exactly cancels out the one that they find themselves in when they are placed near to a magnet. This is exactly the opposite of what happens if we put a piece of pure ferromagnetic iron (that is not a permanent magnet itself) next to a magnet. In that case the iron really likes to encourage magnetic field to concentrate inside it and so it will always be attracted to the magnet whichever way round we place it. The upshot of all this is that the repulsion between a magnet and a superconductor is much more stable than the repulsion between two permanent magnets. A superconductor floating above a magnet never wants to flip over like a permanent magnet does. Great news for wannabe hoverboarders!

In fact, real superconductors have another trick up their sleeves that makes them work even better, and leads to some pretty remarkable effects. If we are clever about how we engineer the materials and if we cool them down in the right way, we can get them to want to stay at a fixed distance from the magnet. Not only does the repulsive force increase if we try to push the superconductor closer to the magnet, but there is an attractive force if we try to pull it away. The superconductor is locked in position relative to the magnet even though it is not physically touching it. If we initially set up an experiment with the magnet resting on the bench and the superconductor floating above it, we can even lift up the superconductor, and as shown in Fig. 4.1, the magnet will hang in mid-air underneath! We can also use this strange effect to get the superconductor to follow a magnetic track. As the track turns a corner the superconductor goes with it because it is magnetically locked in place—it does not want to move out of the field sideways any more than it wants to move up or down

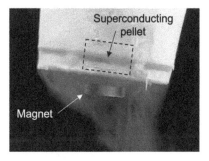

Fig. 4.1: Stable levitation of a magnet suspended below a superconductor.

relative to the track, so it follows the track wherever it goes. This concept has recently been demonstrated in a prototype magnetically levitated (maglev) train capable of carrying 30 passengers at high speeds of up to 620 kph (385 mph) on a 165 m track in Chengdu, China, using high temperature superconductors. Even futuristic hoverboard technology has now been demonstrated. In 2015, a car company that will remain nameless released a video of a real hoverboard floating a few inches above concrete. Although this was a gimmick for an advert, it turns out that it was not just another hoax. The hoverboard really did levitate and it used high temperature superconductors to do it. Nevertheless there was a trick. The necessary magnetic track was hidden beneath a layer of concrete!

Levitation actually has a much broader range of engineering applications than just flying skateboards and fast trains, though these do not capture the imagination in quite the same way. For example, 'frictionless' bearings would be brilliant for all sorts of rotating machines. Things like flywheels (that store energy in a rotating disc) or wind turbines would be considerably more efficient if we could drastically reduce the energy they lose overcoming friction.

So now we have met the basic idea of magnetic levitation and have seen why it might be useful in real applications, but what properties does the superconducting material need to have to be a good levitator, and what materials science tricks can we use to achieve these properties in practice? To answer these questions we will start by introducing the single most important idea in applied superconductivity—the concept of *flux pinning*.

4.1   Flux pinning

To appreciate the full potential of superconductors for levitation applications—and in fact all high current applications—we need to remind ourselves of what actually happens inside a practical superconductor when we put it in a magnetic field. We discussed in Section 3.7 the idea that type I superconductors expel all of the magnetic field they are exposed to, up to some maximum field. If you apply too high a field they suddenly lose superconductivity and become normal conductors again. These particular superconductors are not actually very useful because the critical field at which superconductivity is lost is typically very low—not much higher than the field produced by a standard fridge magnet.

Type II superconductors are much more promising for real applications because they can ultimately withstand much higher magnetic fields. In fact, the high temperature superconductors that are so handy for levitation applications still superconduct in gigantic magnetic fields. But, as we saw in Section 3.7, they do not actually manage to expel all of that field. Instead, some of the magnetic field is allowed to penetrate through an array of 'tubes' called flux lines, each of which carries a single quantum of magnetic flux. In the narrow cylindrical core of each flux line the material becomes normal (non-superconducting), but the rest of the material remains in the superconducting state. We call this the mixed state because the superconducting and normal phases coexist.

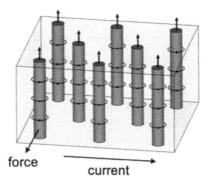

force   current

Fig. 4.2: The flux line lattice in a type II superconductor in the mixed state.

Figure 4.2 shows a schematic diagram of a superconductor in the mixed state, sitting in an external magnetic field pointing vertically upwards. Let's imagine what would happen if we pass a current through the superconductor from left to right. We saw in Section 2.4 that if you put a wire carrying a current into an external magnetic field it will experience a force (unless the current and field happen to be pointing in exactly the same direction). This *Lorentz force* will act in a direction that it is at right angles to both the current and the field—in this case towards us out of the page.[1] This means that as soon as we start passing a current through the superconductor, there will be a force acting on the flux lines. If there is nothing holding them in place, the flux lines will start to move towards us. This turns out to be a major headache because energy is dissipated when flux lines move and the zero-resistance property of the superconductor is lost.

The simplest way of understanding why moving flux lines dissipate energy is by re-membering that they have cores of normal (non-superconducting) material containing normal (non-superconducting) electrons. These normal electrons experience resistance to their motion, just as they do in a conventional metal. The upshot is that if we want to pass large resistance-less currents through a superconductor, we absolutely must do everything we can to stop the flux lines from moving. We do this by deliberately introducing defects (imperfections) into the material to *pin* the flux lines in place. The more effective these defects are, the stronger the pinning force they can exert on the flux lines and the larger the Lorentz force they can withstand before they move and generate resistance. Since the Lorentz force increases with current, there will be an upper limit to the amount of resistance-less current we can pass through a particular cross-sectional area of type II superconductor. The critical current density in type II superconductors is actually defined by the condition where the Lorentz force exactly balances the pinning force produced by the defects. Below this critical current density, the pinning force is larger than the Lorentz force and the flux lines are pinned and cannot move; above this current density the Lorentz force exceeds the pinning force

[1]You can use Fleming's left-hand rule to figure out which direction along the axis the force will point.

and the flux lines start to move generating resistance. Thermodynamically speaking, the material is still in the superconducting state above the critical current density in a type II superconductor, but the zero-resistance property will be lost.

The consequence of all this is that the *microstructure* of the material —what it is like inside on the micrometre and nanometre scale—makes an enormous difference to how much current the superconductor can carry. We can increase the critical current density by packing loads of defects into the superconductor. In contrast, critical temperature and critical field are more fundamental thermodynamic properties of the superconducting phase. They depend on chemistry and external things like pressure that can change the bond lengths for example, but not directly on microstructure. For this reason we sometimes refer to critical temperature and critical field as *intrinsic* properties of the material, whereas critical current density is an *extrinsic* property. Achieving strong flux pinning is by far the most important factor for most practical superconductors because without flux pinning they cannot carry any zero resistance current at all!

Pinning centres

Next we will explore how imperfections in the material actually manage to pin flux lines in place. The most straightforward kind of defect to think about are small bits of non-superconducting material distributed throughout the the superconductor. Materials scientists call them *secondary phase particles* or *inclusions*. Here it is worth mentioning that the word 'particle' is often used in physics to describe things as small as atoms or even subatomic species like electrons. The term is much more general though, and here we are talking not about an individual impurity atom, but a much larger volume containing many thousands of atoms of something chemically different to the superconductor. So how do these 'particles' pin flux lines? Remember from Chapter 3 that the superconducting state is more thermodynamically stable than the normal

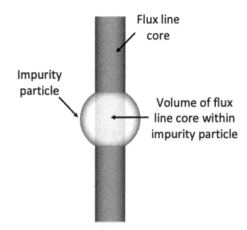

Fig. 4.3: Flux line pinning by an inclusion particle.

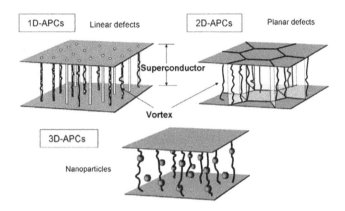

Fig. 4.4: The interaction of flux lines (vortices) with linear, planar and three-dimensional defects. Reproduced from (Namburi *et al.*, 2021) under the Creative Commons Attribution 4.0 International license http://creativecommons.org/licenses/by/4.0/.

state, so creating the normal core of a flux line costs us some energy. The larger the volume of the core, the more the energy it costs to make. This means that a flux line will choose to sit at a location where there is already an inclusion particle (Fig. 4.3). It costs less energy to do so because it can take advantage of the inclusion already being non-superconducting, reducing the total volume that needs to be turned normal to create the flux line core: less volume to convert, less energy required. Moreover, to move the flux line away from the inclusion we would need to put in extra energy to create a new bit of normal core. The consequence is that the flux line is held in place—pinned—by the inclusion. The pinned flux line is sitting in an energy minimum. It is a bit like being in a dip surrounded by hills. To move in any direction from where we are, we would have to go uphill and do some work to increase our gravitational potential energy. We refer to any defect that produces this kind of energy minimum for a flux line as a *pinning centre*. The overall pinning force of a material essentially depends on two factors: how 'strong' each pinning centre is (how deep its energy minimum is) and the density of pinning centres (how many of them there are per unit volume).

Non-superconducting inclusions are by no means the only kind of imperfection in the material that can pin flux lines. There is a whole zoo of other kinds of defect that we can get in crystalline materials, and nearly all of them will provide some sort of pinning. Some act as individual pinning centres like the inclusions described above, and some act collectively. Some are essentially spherical meaning they are just as good with the flux lines pointing in any direction. Others are two-dimensional in shape and yet more are one-dimensional line-shaped defects (see Fig. 4.4). The most effective pinning centres have sizes that are roughly the same as the diameter of the normal core of the flux lines and are distributed with about the same spacing as the flux lines. Flux line diameter

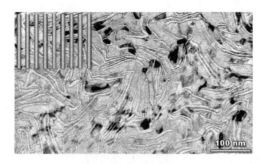

Fig. 4.5: Transmission Electron Micrograph (TEM) of the two phase microstructure of NbTi compared with the flux line scale at 5 T and 4 K. The bright regions are non-superconducting Ti and the medium grey regions are superconducting NbTi. Reproduced with permission from (Lee, 1999).

actually varies pretty widely between different kinds of superconductor, and is governed by a fundamental parameter called the *coherence length* that we will come back to in Chapter 6. Conventional low temperature superconductor (LTS) like NbTi have flux lines around 5–50 nanometres (nm) in diameter, whereas typical high temperature superconductors (HTS) have much smaller diameters of around 1 nm. This means that even very small defects like individual atoms in the wrong place can act as reasonable pinning sites in HTS materials, but larger particles are needed for LTS materials. Also, it's worth remembering that the flux lines get closer and closer together as we increase magnetic field, so ideally if we want to maximise flux pinning we should also tailor the density of defects to the field we want to operate the material in. Figure 4.5 shows the microstructure of NbTi optimised for operation in a magnetic field of around 5 tesla. What you can see in the image are bright regions of non-superconducting titanium finely distributed amongst the mid-grey NbTi superconductor. The inclusions are in the form of thin plates because of the way in which the wire has been processed which you will find out all about in Chapter 5.

Crystal defects

Understanding defects, and how they influence macroscopic properties, is at the heart of all materials science, not just superconductors. The defects that we are going to talk about now are the very same defects that influence the mechanical properties of materials, their optical properties, their corrosion resistance and so on. In particular we are going to talk about crystal defects. You may think of crystals as the pretty, brightly coloured objects with interesting shapes on display in museums, or the bright blue copper sulfate crystals you may have grown at school, or even the diamonds in expensive rings! But, as it happens, the vast majority of all materials, including metals, are crystalline. By this we mean that if we could zoom in and look at them at really high magnification, we would see that the atoms are arranged in a regular, repeating pattern. There are lots of different ways in which this regular arrangement can be disrupted, and we call these imperfections crystal defects.

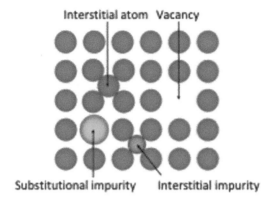

Fig. 4.6: Common point defects in a crystalline material.

Let's start by looking at *point defects* where individual atoms are out of place (Fig. 4.6). As an example, consider an elemental metal like pure niobium that has only one kind of atom in the crystal. There are several different kinds of point defects that can (and do) occur even in this simple system. The first imperfection that we might have is a missing atom—a *vacancy*. This is just somewhere where we should ideally have a niobium atom but it is not there in practice. The second thing that can happen is that an extra niobium atom gets squashed into a location where there should be a gap. This is referred to as an *interstitial* defect. Vacancies and interstitials often occur in pairs because if a niobium atom is pushed out of its proper position and into one of the gaps, it leaves behind a vacancy. Another thing that might happen is that a rogue atom may find its way into the crystal, either masquerading as a niobium atom and sitting in the position where a niobium atom should be, or as an interstitial occupying one of the gaps between the niobium atoms. In more complicated crystals with multiple elements, like YBCO, other types of point defect can also happen—like Y and Ba atoms switching places with one another.

A very important defect in crystals, particularly metals, is something called a *dislocation*. This is not a point defect because it extends over many atoms. The easiest kind of dislocation to visualise is an *edge dislocation*. First of all, think of a crystal as being a stack of sheets, each an atom thick. We call them *atomic crystal planes* or *monolayers*. We could imagine this as being like a ream of paper. If somewhere in the middle of the stack, one of the pieces of paper is actually half a sheet rather than a whole sheet, there will be a defect inside running right across the stack of paper at the edge of the smaller piece of paper. A few sheets above and below it will be bent in the vicinity to try to close up the gap. This line along the edge of the half-sheet is the analogue of an edge dislocation in a crystal. Unlike in the stack of paper, where the dislocation is fixed by where in the stack we inserted the half sheet of paper, in a crystal the extra atomic half-plane can actually move up and downwards in the stack if we apply the right kind of force to it. It is the motion of dislocations that is responsible for plastic (permanent) deformation in metals when they experience a large enough force.

The Wider View

Dislocation pinning

As we will discuss in Chapter 5, metals start to deform plastically—they yield—when dislocations start to move. Like flux lines, dislocations are linear defects, and there is a rather nice analogy between flux line pinning and dislocation pinning. In an edge dislocation, shown schematically in Fig. 4.7(a), there is essentially an extra sheet of atoms in the top half of the crystal (above the dislocation) compared to the bottom half of the crystal. The dislocation itself is actually a line going in and out of the page at the position where the extra plane suddenly stops and is indicated by the inverted 'T' symbol.

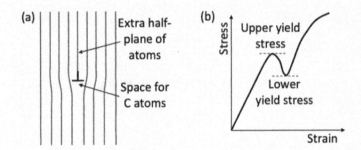

Fig. 4.7: Tensile strain region around a dislocation and the effect of dislocation pinning on the yield stress in steel.

Just as it's possible to increase the performance of superconductors by introducing defects that pin magnetic flux lines, so we can also tailor the microstructure of structural materials to pin dislocations and increase their yield stress. Let's take the example of mild steel, which is basically iron atoms with some carbon atoms as an impurity. Because carbon atoms are small, they can squeeze between the iron atoms and whizz around pretty easily inside the material. But as the carbon atoms are a bit too big for the natural gaps (interstices) in between the iron atoms, there is some compression in their vicinity which leads to a strain energy. Near the core of the dislocation, in the region below the extra half-sheet of atoms, there is a nice zone where the atoms are a bit further apart than they are elsewhere. If the carbon atoms choose to sit at these locations, there is more space, so they do not have to push on the iron atoms so much and the total energy of the system is lowered. So, carbon atoms tend to move to the dislocation cores and additional energy has to be put in to pull the dislocations free. This means that steel has a higher yield stress than pure iron—an effect called *solid solution strengthening*—and as dislocations can move more easily once they have broken free from the carbon atoms, the stress actually decreases after the initial yield in steels (Fig. 4.7(b)).

The final defect that we are going to mention only briefly now is called a *grain bound-ary*. They turn out to be such important defects in superconductors that they have their own chapter in this book! Basically, the diamonds in expensive rings are single crystals. Everywhere inside the material the atoms are lined up in exactly the same way. In most materials, there are in fact lots of different crystals. Within each crys-tal—or *grain*—the atomic planes are lined up with each other, but in the neighbouring crystal the atomic planes line up in a different direction. The wall between the crys-tals is called a grain boundary. These walls are relatively thin but have a large area, so they are two-dimensional defects. This means that they are particularly good at flux pinning in superconductors—there is a good chance that flux lines can line up along the boundaries and save themselves a decent amount of energy. In fact, in many low temperature superconductors including niobium-tin and magnesium diboride dis-cussed in Chapter 8, they are responsible for the majority of the flux pinning. This means that making the material with small grains to increase the number of grain boundaries will improve performance.

If you want a fun way to see what crystal defects look like, they are reproduced rather nicely in ears of corn on the cob! Inspect them closely and you will soon see some imperfections in the regular arrangement of the kernels. See if you can spot a missing kernel—a vacancy, or a kernel that is a different size to the rest—an impurity atom. Maybe you can see a really small kernel squeezed in where it should not be—an interstitial. Also look at what happens in the surrounding area—the defects will have an effect even a few kernels away. If you are lucky, you may even be able to spot a defect that looks like an edge dislocation where a row of kernels suddenly stops.

4.2 Stable levitation

So, we have seen that flux pinning is needed to get high critical current densities in superconductors. As it turns out, strong flux pinning really saves the day when it comes to levitation as well. Returning to the scenario where we place of piece of superconductor above a permanent magnet. When it is a long way away from the magnet, the superconductor will be out of the magnetic field. As we slowly move it towards the magnet, circulating surface currents will be set up in the superconductor to stop the magnetic field from penetrating to the interior (the Meissner effect). When the superconductor is close enough to the magnet that the field exceeds the lower critical field, we would expect that flux lines would be created inside the superconductor, but in practice the flux lines must physically make their way into the superconductor from the surfaces. If we have a material with strong flux pinning, the flux lines get stuck as soon as they enter the superconductor and they find it really difficult to get in. There is a traffic jam! This means that a larger magnetic field can be screened and the repulsive force experienced is significantly stronger than we would expect for a superconducting material with no flux pinning. This is really good news for applications!

In fact, in addition to increasing the repulsive force, we can also exploit the trapping of magnetic flux to make levitation much more stable. Instead of moving a superconductor close to a magnet after it has been cooled into the superconducting state, we can do a slightly different experiment called *field cooling*. As the name suggests, this involves cooling the superconductor in the presence of a magnetic field. Imagine starting with the superconductor warm (above its critical temperature) and resting on a non-magnetic spacer block on top of the magnet.[2] Since the superconductor is initially in the non-magnetic normal state, the field from the magnet would penetrate straight through it without being expelled (see Fig. 4.8). If we then cool the material down into its superconducting state, if the superconductor has strong flux pinning, instead of the field being expelled it will be trapped inside the material. We can remove the spacer and the superconductor will stay in the same place, floating above the magnet. We would have to do work to move it closer to the permanent magnet because the magnetic field is stronger there and we would need to forcibly push extra flux lines in from the edge. Importantly though, we would also have to do work (put in energy) to pull the superconductor away from the magnet because the flux lines cannot easily get out. In fact, if we do forcibly remove the superconductor from the magnet, provided we keep it cold of course, the flux lines stay trapped inside and we have turned it into a permanent magnet. It stays magnetised even when the external magnetic field has been removed. We can replace it at a later point and it will click back into its favourite position floating at pretty much the same height above the magnet.

---

[2] A spacer separates two other objects at a particular distance.

Under the Lens

Estimating levitation force

To get an estimate for how big the levitation force is between a superconductor and a permanent magnet we first need to have an equation that tells us the force on a magnetic material in a magnetic field. The magnetic work done in increasing the magnetisation of a material by an infinitesimally small amount dM is $dW = VBdM$ where V is the sample volume and B is the applied magnetic field and the magnetisation M is the sample's magnetic moment per unit volume (see Appendix C). Now think back to the standard definition of work done: force multiplied by the distance moved in the direction of the force. The magnetic work done in moving a magnetic material by an infinitesimal distance (dz) is $dW = F_z dz$ where F_z is the z-component of the force acting on the magnetic sample. Therefore,

$$F_z = \frac{dW}{dz} = \frac{d}{dz}(VBM) \tag{4.1}$$

If the sample is a conventional magnet (a ferromagnet), its magnetisation is pretty much independent of the applied field so the force will just be $F_z = VM\frac{dB}{dz}$. The field gradient $\frac{dB}{dz}$ is negative because the field drops off with increasing distance between the magnets, z, so the force will be negative (attractive). On the other hand, if the sample is a superconductor the magnetisation is negative (in the opposite direction to the applied field) so the force will be positive (repulsive). But for a superconductor in the Meissner state, we know that its magnetisation is not constant—it depends on the applied magnetic field, according to the equation $M = -\frac{B}{\mu_0}$, where μ_0 is the permeability of free space. Therefore the levitation force depends on both the strength of the magnetic field and the gradient of the magnetic field $\frac{dB}{dz}$.

$$F_z = -\frac{V}{\mu_0}\frac{d}{dz}(B^2) = -\frac{2V}{\mu_0}B\frac{dB}{dz} \tag{4.2}$$

In this case, since $\frac{dB}{dz}$ is negative, the force always turns out to be positive—a superconductor in its Meissner state is repelled from the magnet. We can maximise the repulsive force by maximising both the magnetic field and the magnetic field gradient $(\frac{dB}{dz})$ produced by the permanent magnet. Taking rough values for the magnetic field and field gradient produced by a strong neodymium iron boride magnet of 1 T and -20 T m^{-1}, respectively, we find that the force per unit volume of superconductor is about 30 N per cm^3 of superconductor when it is placed on the surface of the magnet. Since the density of $YBa_2Cu_3O_7$ HTS is about 6 g cm^{-3}, the downwards force due to gravity per cm^3 is much smaller, at only 0.06 N. The superconductor will therefore be pushed upwards, away from the magnet until the levitation force has decreased sufficiently to exactly balance the gravitational force.

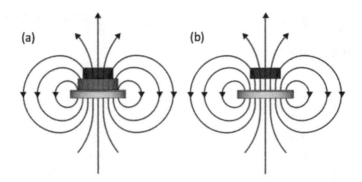

Fig. 4.8: Field cooling of a superconducting pellet, (a) before and (b) after the insu-lating spacer block is removed.

4.3 Trapped field magnets

We have seen that superconductors are better than normal ferromagnetic materials like neodymium iron boride ($Nd_2Fe_{14}B$) for levitation, but can they also be better permanent magnets? The quick answer is yes... and no. It really depends what we mean by better. If we want to produce stronger magnetic fields, then superconductors win hands down. The world record magnetic field produced by a superconducting pellet (usually referred to as a *bulk*) is over 17 tesla. This is considerably more than 10 times the magnetic field produced by the strongest ferromagnets. But of course superconductors need to be cooled down which restricts which applications they can be used for. More than that, as we have already seen, just cooling it into the superconducting state does not make the superconductor magnetic. We have to get flux lines trapped inside to magnetise it.

In fact, we also need to magnetise conventional ferromagnetic materials to make them into permanent magnets. This is because, although the atoms in ferromagnets have permanent magnetic dipole moments which are strong enough that they spontaneously line up with their neighbours, the materials split into domains that are randomly aligned. The consequence of this is that we need to apply an external magnetic field to both superconductors and ferromagnetic materials to turn them into permanent magnets. In a superconductor this is to get the flux lines trapped inside, whereas for a ferromagnet the purpose is to force all of the domains to line up with each other. The more difficult it is for the flux lines to get out again, or the ferromagnetic domains to misalign themselves again, the better the permanent magnet will be. We call ferromagnets that do not easily demagnetise *hard ferromagnets*, and so by analogy, we call superconductors with strong flux pinning *hard superconductors*. Before we get on to how we do the magnetisation process in practice, let's think a little more about what the maximum possible magnetic field produced by a bulk superconducting magnet would be.

Start by imagining a hollow superconducting tube rather than a solid cylindrical pellet. If we somehow manage to get currents to whizz around its circumference—do not worry how, for the time being—they will produce a magnetic field inside the tube pointing along its axis (Fig. 4.9). This is pretty much exactly the same as what happens in the solenoid magnet we talked about in Section 2.1 except that the currents are not confined to circulating in wires this time. We know that in a solenoid the magnetic field generated increases as the current in the wire increases, so we will get the biggest possible field inside the tube if the biggest possible current is circulating. But we also know that the maximum current density that can be sustained in a superconductor without resistance is the critical current density of the material. So to get the maximum field inside the tube, currents must circulate at the critical current density.

A solid cylindrical pellet can be thought of as a nest of concentric tubes with the critical current density circulating around each one. To work out the total magnetic field produced in the centre of the pellet, we just have to add up the field produced by each separate tube. The total magnetic field gradually decreases as you go radially outwards from the centre of the cylindrical pellet to the edge. This is because the magnetic field generated by a particular tube is confined to the region inside it, if the tube is sufficiently long, so the innermost tubes with a small radius tubes will only contribute to the field near the centre of the pellet. When we get to the outside curved edge of the superconductor we are outside all of the nested tubes, and the magnetic field will be zero. So we find that the magnetic field produced by a bulk superconductor is largest in the centre and gradually decreases to zero at the curved edge if current is circulating uniformly at the critical current density throughout.[3] The larger the critical current density, the larger the field in the centre will be. Moreover, because we can always boost the central magnetic field by adding an extra larger radius tube around the outside, it is easy to see that the magnetic field in the centre of a superconducting bulk scales with its radius. The wider the cylindrical pellet, the higher the magnetic

Fig. 4.9: Magnetic field created by a tube of superconductor with a current circulating around the curved surface and the model of a bulk superconducting pellet as a nest of concentric tubes.

---

[3]This assumes the ideal situation where the cylindrical pellet is infinitely long. In practice, the pellets are likely to have similar height and diameter, so the field distribution will not be quite the same as this simple model.

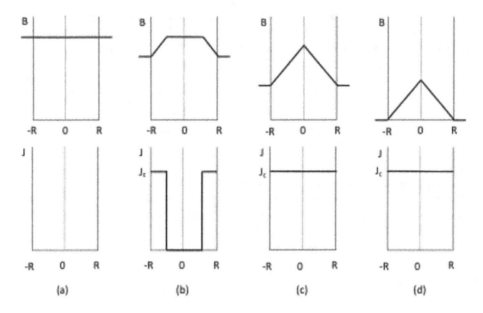

Fig. 4.10: Evolution of the distribution of magnetic field and the magnitude of the screening current density during removal of external field after field cooling.

field it produces will be. Interestingly, the same thing is not true in a conventional ferromagnet. In that case, because the magnetism is produced by atomic magnetic dipoles that are evenly distributed throughout the material, increasing the size of the ferromagnet will not change the magnetic field it produces. The field produced by a ferromagnet is essentially fixed by how big its atomic moments are, whereas in a superconductor we can get bigger fields by both engineering the material to have higher critical current density and by making the bulks larger in radius.

One final, rather complicated argument is needed to see how this maximum field case relates to the trapped field you end up with when you field cool a superconductor and then remove the external field. In fact, it turns out that the two situations lead to identical current distributions. If we cool a superconductor with strong flux pinning in an external magnetic field, the magnetic field will not be expelled immediately the material enters the superconducting state because the flux lines are trapped by the pinning centres and cannot get out (Fig. 4.10a). When we start reducing the external magnetic field, resistance-less currents begin to circulate around the surface of the superconductor to boost the internal magnetic field, thus shielding the interior of the sample where there is trapped flux from the decrease in external field (Fig. 4.10b). Since ideally the superconductor wants to shield as much of its volume as possible from the changing field, the currents will circulate in as thin a surface layer as they can by flowing at the maximum possible current density—the critical current density. Flux lines can then start to move out of the edges of the superconductor from this surface shell where the critical current is flowing. It is a bit like opening a gate—in regions carrying the critical current density, the flux lines can start to move and some of them are let out of the edge of the sample. So there is a layer of material at the outside edge of the superconductor that is carrying screening currents at the full critical current density and from which some magnetic flux lines have partially escaped, but there is a central zone of the sample that still does not have any current circulating and has its original number of magnetic flux lines. This means there is non-uniform field profile within the superconductor, with the edges having lower trapped field than the centre.

As we continue to reduce the external magnetic field, the screening currents have to produce a larger magnetic field to make up for the loss of the externally applied field, and this can only happen by increasing the thickness of the layer through which the currents flow because they are already circulating at their maximum possible current density. Eventually the entire bulk, right to the centre will be carrying circulating currents at the full critical current density value. At this point, there will be a steady gradient in the amount of trapped field increasing from the edge to the centre of the pellet (Fig. 4.10c). Continuing to remove the external magnetic field will allow more and more flux lines to gradually make their way out of the edges of the pellet, but the screening currents and the conical-shaped field profile will remain—even when the external field has been completely removed (Fig. 4.10d). This situation is identical to the maximum field case discussed previously with currents circulating at J_c everywhere.

The idea that in a strongly pinned superconductor, current is either flowing at the full critical current density value or is zero is called the *Bean model*—not because it has any resemblance to the vegetable, but because it was originally proposed by Bean and Livingstone in 1962. Modifications and improvements to this basic model have

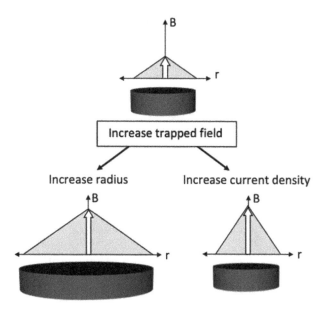

Fig. 4.11: Factors affecting the magnetic field distribution in trapped field magnets.

been made over the years and this family of theories is now collectively called *critical state models*. The important takeaway message is that we can maximise the magnetic field trapped by a bulk superconductor in two ways: by maximising the critical current density of the material (optimising flux pinning) and by increasing the radius of the bulk pellet, as shown in Fig. 4.11.

Under the Lens

Estimating trapped field

In an infinitely long solenoid, the field inside the bore is uniform and has the value $B = \mu_0 n I$ where n is the number of turns per unit length, I is the current in the wire and $\mu_0 = 4\pi \times 10^{-7}$ Hm^{-1} is the permeability of free space. Modelling the bulk superconducting pellet as a nest of concentric thin walled tubes of thickness Δt, we can estimate the magnetic field as a function of radial position in the bulk. We assume here, for simplicity, that each tube is long so we can use the infinite solenoid model.

For a single tube with a total of N turns and with length L, the current density J is the total current in all of the turns of the coil added together, NI, divided by the cross-sectional area through which the current flows $L\Delta t$.

$$J = \frac{NI}{L\Delta t} = \frac{nI}{\Delta t} \tag{4.3}$$

We can then substitute $J\Delta t = nI$ into the infinite solenoid equation to get the field inside a single tube B_{tube} assuming the current is flowing at the critical current density J_c.

$$B_{\text{tube}} = \mu_0 J_c \Delta t \tag{4.4}$$

Now, the number of tubes that contribute to the field at a radial position r is the number of tubes with radius $>r$. If the pellet has a radius R, the number of tubes contributing to the field is given by $\frac{R-r}{\Delta t}$. Since each of these tubes contributes B_{tube} we can write down a simple equation for the field as a function of radial position inside the superconductor.

$$B_r = \mu_0 J_c \Delta t \left(\frac{R-r}{\Delta t} \right) = \mu_0 J_c (R - r) \tag{4.5}$$

The maximum field is at the centre when $r = 0$ is $B_{\text{peak}} = \mu_0 J_c R$. Therefore, we can see that to increase the peak trapped field we can either increase the radius of the sample or increase J_c. It is also worth noting that the magnetic field decreases linearly with r and that J_c scales with the magnitude of the gradient of the magnetic field in the radial direction.

4.4   Practical magnetisation methods

So let's return to thinking about how we start these currents circulating. We have already discussed the field cooling method that involves cooling down the superconductor in the presence of a magnet and then taking away the magnet. The problem is that the field trapped in the superconductor is limited to the field of the magnet that you are magnetising it with. To get large trapped fields we need to use a high field electromagnet—probably an expensive superconducting one—which begs the question: Why bother with the bulk superconductor at all? Why not just use the electromagnet itself? Moreover, how do we get the bulk out of the electromagnet and into the device where we want to use it without warming it up? An alternative method to magnetise a bulk superconductor, involving cooling in zero field and then applying an external field and removing it again, is even worse; in that case you actually need to apply twice as high a magnetic field as you want to trap in your magnet! This is the crux of why bulk superconductors are not yet used in more applications.

Recent attention has been focused on developing another, more practical method called *pulsed field charging*. This uses a smaller and cheaper conventional copper electromagnet that can be easily taken away after it has been used, leaving the magnetised bulk accessible for the intended application. It works by cooling the superconductor in zero field and then, as the name suggests, applying a rapid field pulse via an electromagnet. Because the field ramps up very quickly, the flux lines are rapidly forced into the superconductor. As there is resistance to moving flux lines, considerable heat is generated, which raises the temperature of the superconductor making it easier for more flux lines to get in. This pulsed field magnetisation method is definitely promising, but it is still in the relatively early stages of development, and the trapped fields achieved are quite a long way below the world record results obtained by field-cooling.

4.5   Applications of trapped field magnets

We have talked about the difficulties with magnetising bulk superconductors, but what advantages do they offer over alternative technologies? Permanent magnets are useful because of their compact size compared to wound magnets. The considerably higher magnetic fields that superconducting bulk magnets can offer over ferromagnets make them attractive for applications that need to be small or lightweight like portable or desktop systems. For example, imagine a small, cheap MRI machine that could be in every doctor's surgery for imaging fingers, elbows and knees. Or maybe you need a portable magnetic field for detecting cracks in aircraft. Bulk superconducting magnets would be really good for this. What about electric motors? As it happens, electrical motors and their systems use more energy than anything else, accounting for a whopping 43–46 % of global electricity consumption and producing a mammoth 6 billion tonnes of CO_2 (6×10^{12} kg). Superconducting bulks are potentially really useful because of the higher efficiencies compared to conventional motors as well as the improved power-to-weight ratio which is particularly important for aircraft applications. Since a motor typically uses a magnetic coil and a permanent magnet to generate a turning force, it may even be possible to magnetise a superconducting bulk (that replaces the permanent magnet) in situ using the electromagnet winding itself.

The Wider View

Magnetic drug targeting

One of the difficulties about conventional drug therapies where the drug is injected into the bloodstream or taken orally, is that it is not delivered to the specific location where it is needed. This dilution throughout the body means that a larger dose of the drug is required and this can lead to unpleasant side effects. An example that is probably familiar is chemotherapy for cancer treatment. It would be fantastic if the drug could be targeted directly at the tumour itself. One of many possible ways of doing this is to enclose the drug inside mini bubbles that are tagged with magnetic nanoparticles. A magnet held outside the body can then be used to trap the bubbles as they move past the tumour (Fig. 4.12).

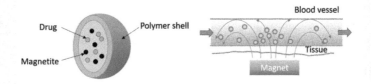

Fig. 4.12: Magnetic drug targeting.

The magnetic nanoparticles that are used are *superparamagnetic* materials. Paramagnets are magnetic materials in which the atoms are permanent mini magnets. When an external field is applied, they start to line up with it, and their magnetisation increases. The 'super' bit just means that they produce a larger magnetisation than normal paramagnets. Ferromagnets would be stronger magnets, but we are not allowed to inject them into the human body. The strength of the attractive force depends on the strength of the magnetisation of the magnetic nanoparticle, which in this case is roughly proportional to the applied field. So increasing the field strength of the external magnet will increase the chance of trapping the bubble containing the drug. Increasing the field gradient—how quickly the field drops off from the top surface of the magnet—also increases the trapping force.

The main difficulty with this technique is that the magnetic forces that can be exerted by conventional magnets on the nanoparticles flowing in blood vessels some distance away are quite small and effective trapping can only be achieved for relatively small diameter blood vessels near the skin. Replacing the external magnet with a superconducting trapped field magnet could enable higher fields to be generated which may enable magnetic drug targeting in larger blood vessels where the flow is faster and blood vessels that are located nearer the centre of the body.

Chapter summary

- Magnetic levitation using conventional permanent magnets alone is always unstable, but we can get much better stability using superconductors.

- Flux pinning is the single most important concept in applied superconductivity. This is because it is absolutely essential for all high current applications including levitation and trapped field magnets.

- Strong flux pinning enables stable levitation because flux lines are trapped inside the superconductor, effectively locking it at a certain height relative to the magnet.

- Bulk superconductors can be turned into permanent trapped field magnets if they have strong flux pinning. The field produced is highest in the centre and drops off towards the edge of the superconducting pellet. The maximum trapped field can be increased by increasing the radius of the pellet and by increasing the critical current density of the material by increasing flux pinning strength.

- Structural defects in crystals, such as point defects, dislocations and grain boundaries, are not only necessary for flux pinning in superconductors, but are universally important in all materials. Understanding how to control defects like these to optimise performance is what materials science is all about!

5

NbTi: The Wonderful Workhorse

Niobium-titanium (NbTi) does not, at first sight, appear to be a particularly special superconductor. It does not have a high critical temperature and it cannot withstand super high magnetic fields. It does not have weird chemistry or exhibit an exotic flavour of superconductivity. Nevertheless, it is the single most important superconducting material we have. It is used in all of the MRI magnets worldwide. It is what is used in the magnets that bend the proton beams in the Large Hadron Collider (LHC) at CERN. In fact, the vast majority of commercial large scale applications of superconductors use NbTi. Even if ultra high fields are required, NbTi magnets are typically used to generate the background field and coils made from 'better' superconductors are placed inside to boost the field. So why is it so used so ubiquitously? It has one major selling point that the other technological superconductors can only dream of, and it has nothing to do with its superconducting properties. The material is *ductile*, which means that we can permanently change its shape without it breaking. This is super important because we need to be able to make many kilometres of the stuff as thin wires. We also need to be able to wind them into magnets, and the product needs to be mechanically robust. Amazingly, as we will discuss in Chapter 8, it is also possible to make wires out of brittle superconductors, but the processing is much more expensive and difficult. NbTi is nearly always the answer to the question: Which superconductor should I use? For this reason, it is often referred to as *the workhorse superconductor*, but as you will see in this chapter, it is by no means a boring material. In fact it is an exquisite example of top class materials science.

5.1 The Nb-Ti phase diagram

Before we get into the nitty-gritty of how to optimise the performance of NbTi wires, first we need to get to grips with the chemistry of the Nb-Ti alloy system. One of the most important tools that materials scientists rely on for understanding a chemical system is the phase diagram. The term *phase* relates to how the atoms or molecules are arranged within the material, and phase diagrams show what chemical phase (or phases) are most energetically stable under certain conditions. For simple systems like water, we saw in Section 3.2 that we generally only need to consider three different phases relating to the three different states of matter—solid, liquid and gas.[1] However, in general it is possible for a system in a given state to exist as several different phases.

---

[1] However, there are multiple different structures of ice, each of which is formally a different phase.

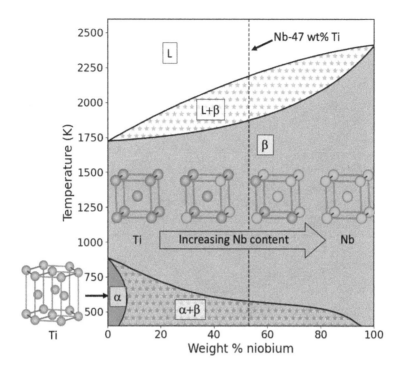

Fig. 5.1: Nb-Ti phase diagram. Insets show the crystal structures of the α-Ti and β-NbTi phases with Ti atoms shown in blue and Nb atoms shown in green.

For example, there is a whole zoo of different solid forms of carbon including graphite, diamond, buckyballs, nano-tubes and graphene. Each of these allotropes[2] is a different solid phase because the carbon atoms are arranged differently, resulting in them having different physical properties.

Water and carbon are both examples of *single component systems*. These are systems that consist of a single chemical substance. This substance could be a pure element made up of one type of atom (like the carbon case) or could be a compound consisting of small molecules (like the H_2O molecules in water), which do not break up into their constituent elements when melted or boiled. Phase diagrams get more complicated when we start looking at systems with multiple components, because each additional component adds an extra dimension to the phase diagram. Niobium-titanium is an example of a binary system because it has two components: niobium and titanium atoms. Fortunately, since we are only interested in standard room pressure (1 atmosphere (atm) = 100 kilopascals), we can draw a useful two-dimensional diagram with alloy composition on the x-axis and temperature on the y-axis, as shown in Fig. 5.1. At high temperature, the liquid phase (L) extends across the whole composition range.

---

[2]Allotropes are different solid forms of a pure element. More generally, for materials that are not chemical elements we use the term *polymorph* instead.

This is not surprising because we would expect to be able to melt alloys of any composition if we go to high enough temperature. The melting point of pure niobium is higher than that of pure titanium, so the lower edge of the liquid phase field is higher at the Nb-rich end of the diagram.

Now cast your eye downwards on the diagram, ignoring for the moment the region labelled 'L + β', until you get to the region labelled 'β'. Here the greek letter 'β' (beta) refers to one of the solid phases that is present in the Nb-Ti system.[3] Like the liquid phase, the β phase region extends across the whole width of the phase diagram—all the way from pure titanium to pure niobium. This essentially means that the atoms inside the material are arranged in the same way regardless of how many of them are Nb atoms and how many are Ti atoms. Materials with atoms arranged locally in a regular, repeating pattern are described as being *crystalline*. Nearly all materials are naturally crystalline, including metals and alloys like niobium-titanium. If the basic repeating unit of the pattern—its *unit cell*—is the shape of a cube, the material is said to have a cubic crystal structure. There are seven different crystal systems in total, each with a different unit cell shape.

It turns out that the β phase of NbTi has the *body-centred cubic (BCC)* crystal structure shown schematically in the inset of Fig. 5.1. At the titanium side of the phase diagram, you can think of the structure as consisting of a regular array of cubic unit cells with a Ti atom at each corner and right in the centre of the cube (at the position referred to as the body-centre). As we move across to the right on the phase diagram, we gradually replace some of the Ti atoms, at random, with Nb atoms. It is possible to carry on replacing Ti atoms with Nb atoms until there are no Ti atoms left (pure Nb) and the crystal still has the same body-centred cubic structure. The β phase is an example of something called a *substitutional solid solution* because we are essentially dissolving Nb in Ti by substituting Nb for Ti atoms (or vice versa). Because the β phase extends across the whole width of the phase diagram, Nb-Ti is an example of a system showing *complete solid solubility* at elevated temperatures. There is a set of empirical rules, known as the Hume-Rothery rules, that can help us to predict how soluble one element will be in another.[4] It turns out that elements tend to show good solid solubility over a wide composition range if, like titanium and niobium, the atoms have a similar size and similar chemistry (i.e. are close to each other on the periodic table).

In between the β phase region and the liquid phase region, there is a region labelled 'L + β' where both the liquid and β phases coexist under equilibrium conditions. You will probably be aware that a pure substance (like elemental Nb or Ti at the extreme edges of the phase diagram) melts at a single temperature—the melting point. However, impure substances melt over a range of temperatures, leading to the opening of this two phase region on the phase diagram. The upper boundary of the $L + \beta$ region is called the *liquidus* line and represents the temperature at which the liquid of a certain

---

[3] By convention we refer to different solid phases in a system by greek letters to distinguish them from chemical elements.

[4] William Hume-Rothery founded the Department of Metallurgy (now the Department of Materials) at the University of Oxford in the 1950s.

The Wider View

Crystal systems

There are seven different crystal systems, each with different symmetries. The unit cells are defined using three lengths called the lattice parameters (a, b and c) and three angles (α, β and γ). By convention, α is the angle between the sides b and c, β is the angle between sides c and a and γ is the angle between a and b.

Table 5.1: Unit cells of the seven crystal systems.

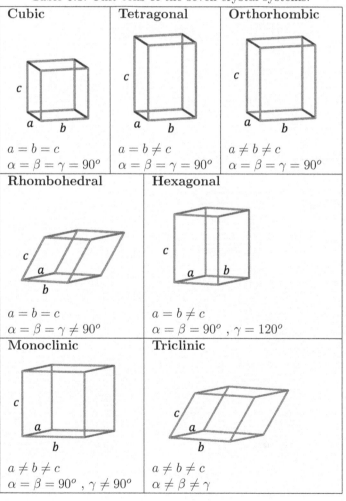

Cubic	Tetragonal	Orthorhombic
$a = b = c$ $\alpha = \beta = \gamma = 90^o$	$a = b \neq c$ $\alpha = \beta = \gamma = 90^o$	$a \neq b \neq c$ $\alpha = \beta = \gamma = 90^o$

Rhombohedral	Hexagonal
$a = b = c$ $\alpha = \beta = \gamma \neq 90^o$	$a = b \neq c$ $\alpha = \beta = 90^o$, $\gamma = 120^o$

Monoclinic	Triclinic
$a \neq b \neq c$ $\alpha = \beta = 90^o$, $\gamma \neq 90^o$	$a \neq b \neq c$ $\alpha \neq \beta \neq \gamma$

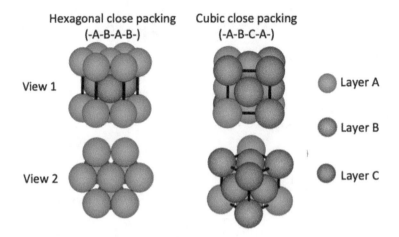

Fig. 5.2: Hexagonal close packed and cubic close packed structures.

alloy composition starts to solidify. The lower boundary is known as the *solidus* and gives the temperature at which the last remaining liquid solidifies.

Now see what happens if we continue to lower the temperature from the β phase region towards room temperature. At the titanium-rich side of the diagram we see a new solid phase labelled 'α' (alpha). In the α phase, the Ti atoms are arranged in a different crystal structure called *hexagonal close-packed* (*HCP*) . This structure can be described as stacks of two-dimensional layers with the atoms arranged in a hexagonal pattern within the layers. If you model the atoms as hard spheres like ping-pong balls, each atom touches six other atoms within the layer. There are two different ways in which these layers can be stacked efficiently on top of each other to produce close-packed structures that minimise empty space. Hexagonal close packing is one of these stacking arrangements, the other being *cubic close packing* (also known as *face-centred cubic* (*FCC*), as shown in Fig. 5.2. The percentage of the volume occupied by atoms as opposed to free space is higher in the close-packed structures than in the body-centred cubic structure of the NbTi β phase. It is, again, possible to replace some of the Ti atoms in the hexagonal close-packed α structure with Nb atoms, but the structure quickly becomes unstable—the close-packed structure simply cannot accommodate very many Nb atoms (which are slightly larger than Ti atoms). Instead, it becomes energetically favourable for the alloy to separate into a mixture of Ti-rich α phase and Nb-rich β phase. Under equilibrium conditions, the compositions of the two phases in the mixture at a certain temperature (i.e. the concentrations of Nb and Ti in each phase) are given by the lines that bound the $\alpha + \beta$ region on the phase diagram—the *solvus* lines. As we cool further, the solid solubilities of niobium in titanium and vice versa change a bit, but we would still expect these same two solid phases to coexist across a large range of alloy compositions.

Under the Lens

The Lever rule

The phase diagram can be used to figure out, not only which phases we would expect to find in a particular alloy composition and temperature, but also how much of each phase should be present under equilibrium conditions in a two phase mixture. The method is as follows, using the $\alpha + \beta$ phase region of NbTi as an example.

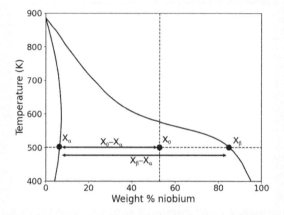

Fig. 5.3: Section of the Nb-Ti phase diagram.

- Draw a horizontal line across the $\alpha + \beta$ at the temperature of interest. This is called a *tie line*.

- Read off the compositions of the two phases: X_α and X_β. These are the positions where the tie line reaches the boundary with the single phase α and β regions either side, respectively. In this example X is given as a weight fraction of Nb, but the same procedure can be followed with atomic fractions if preferred.

- We can then write the overall alloy composition X_0 in terms of the fractions of each phase: f_α and f_β where $f_\alpha + f_\beta = 1$.

$$X_0 = f_\alpha X_\alpha + f_\beta X_\beta = f_\alpha X_\alpha + (1 - f_\alpha)X_\beta \qquad (5.1)$$

- Rearranging the equation we can find expressions for the fractions of each phase.

$$f_\alpha = \frac{X_\beta - X_0}{X_\beta - X_\alpha} f_\beta = \frac{X_0 - X_\alpha}{X_\beta - X_\alpha} \qquad (5.2)$$

This equation is known as the *Lever rule*.

If we take the standard NbTi alloy with 47 weight percent Ti (53 weight percent Nb), we can use the Lever rule to estimate the fraction of α and β phases at a chosen temperature. Let's take 500°C because we can easily read the compositions off the phase diagram: $X_\alpha \approx 0.06$ and $X_\beta \approx 0.85$.

$$f_\beta = \frac{X_0 - X_\alpha}{X_\beta - X_\alpha} = \frac{0.53 - 0.06}{0.85 - 0.06} = \frac{0.47}{0.79} \approx 0.6 \tag{5.3}$$

So under equilibrium conditions at 500°C, we would expect our alloy to contain 60% of the β phase by weight. If we want to know what the volume fraction of each phase is, we would need to know the density of each phase. In practice, the fraction of the superconducting β phase will tend to be higher than this because it is difficult to reach equilibrium.

5.2   Choosing the alloy composition

The beta phase (β) is the superconducting phase we refer to when we talk of niobium-titanium. The alpha phase (α) also superconducts, but because it has a critical temperature below 4 K it is non-superconducting when operating at liquid helium temperature. From the phase diagram, we have seen that at low temperatures (anywhere below about 600°C), we would expect most compositions of NbTi alloy to contain two phases: superconducting β and non-superconducting α. This may at first appear to be a disadvantage for making superconducting wires because the regions of non-superconducting α phase will reduce the cross-section of the wire that can carry superconducting current. On the contrary though, it actually turns out to be very fortunate because the α particles can act as excellent flux pinning centres. The key is to tailor the way we process the material to control how much α phase we end up with and how it is distributed microscopically throughout the alloy.

Consider an alloy with about 50% Nb and 50% Ti. If we heat it for long enough at 1500°C the atoms will move about by a process called *solid state diffusion* until the composition is uniform throughout—it will be single phase β NbTi. Now, if we cool it really slowly to a temperature in the two phase $\alpha + \beta$ region, some of the β phase will transform to α phase. We call the α phase a *precipitate* and the $\beta \rightarrow \alpha$ phase transformation is called *precipitation*. The precipitation process occurs by a two-stage mechanism called *nucleation and growth*. In the nucleation stage, Ti atoms which are constantly jiggling around, sometimes come together and form a cluster. If the cluster is very small, it is most likely that it will simply disappear again. However, if by chance the cluster manages to get large enough we refer to it as a *nucleus* and it becomes energetically more favourable for it to grow in size rather than to shrink. In the subsequent growth stage, Ti atoms diffuse and attach to these α nuclei until the equilibrium phase fractions and compositions are reached. As we will see in Chapter 8 the solid state diffusion process required for the Ti and Nb atoms to redistribute themselves can happen much faster at higher temperatures. This means that if we cool the alloy from the β phase region quickly the atoms do not have time to get to their equilibrium positions. In fact, if we cool fast enough, we can suppress the α

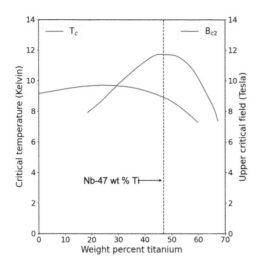

Fig. 5.4: Effect of composition on critical temperature and critical field in Nb-Ti.

precipitation process altogether and keep the whole material in the superconducting β phase. We call this a *supersaturated solid solution* because it is a solid solution that has more of the impurity component (in this case Ti) dissolved in it than is possible according to the equilibrium phase diagram. Supersaturated solid solutions could lower their free energy by separating into two phases, but the atoms are not mobile enough to redistribute themselves.

This means that we can actually make β NbTi alloys with the body-centred cubic crystal structure over a large range of chemical compositions, all the way from pure Nb to alloys with as much as 70% Ti by weight, that do not, in practice, spontaneously decompose into the two separate solid phases at room temperature. This allows us to investigate how the superconducting properties of β NbTi vary with Ti content (see Fig. 5.4). As you can see, the superconducting transition temperature, T_c, does not change very much when we increase titanium content—at least until we near the Ti-rich compositions. In contrast, the upper critical field (B_{c2}) is quite strongly influenced by composition, with a maximum value of around 12 tesla at around 50% by weight of Ti. It is for this reason that the standard alloy composition used for NbTi superconducting wires has 47% titanium by weight (Nb-47 wt % Ti).

Under the Lens

Nucleation theory

Many phase transformations happen by a process called *nucleation and growth*. The nucleation stage involves atoms clustering together. The simplest situation to think about is the formation of a solid phase particle from a liquid. Atoms are in constant motion and they will periodically come together to form a small cluster of solid. If that cluster gets large enough, it will tend to grow further, but if it is too small it will most likely shrink again. This behaviour is governed by the interplay between two different energy terms: volume free energy and surface energy. The first of these is the difference in Gibbs free energy between the solid state and the liquid state and provides the driving force for solidification. Below the melting point, the solid state is more stable, and so the free energy of the solid will be lower than the free energy of the liquid. The amount of free energy saved by a particle solidifying scales with its volume. However, we also have to take into account the fact that the boundary between the solid and the liquid will have an energy cost associated with it that scales with surface area of the particle. The origin of this surface energy can be understood by thinking about the surface as a region where the bonding is not ideal—there are broken bonds. When the particle is very small, the surface area to volume ratio is high and so it costs energy overall and the solid particle will be unstable, but as the particle grows, the volume free energy term starts to dominate, and the particle will become a stable nucleus.

Assuming the growing solid particle is spherical in shape, the total Gibbs free energy change associated with converting a particle of radius r from liquid to solid is given by

$$\Delta G = -\frac{4}{3}\pi r^3 \Delta G_v + 4\pi r^2 \gamma \tag{5.4}$$

where $\frac{4}{3}\pi r^3$ is the volume and $4\pi r^2$ is the surface area of the spherical particle, ΔG_v is the free energy saved by solidification per unit volume and γ is the surface energy per unit area. The first term has a negative sign because it is the energy saving associated with the particle solidifying, whereas the second term is positive because it is the energy cost associated with the surface. This is shown graphically in Fig. 5.5.

The particle will become stable when growing larger starts to decrease its total energy—that is, as soon as it passes the peak of the energy curve. Our standard mathematical method for finding a peak—a turning point—is to find where the first derivative (gradient) of the curve becomes zero. Here that means that we need to differentiate the expression for ΔG as a function of r and set it equal to zero.

$$\frac{d\Delta G}{dr} = -4\pi r^2 \Delta G_v + 8\pi r \gamma = 0$$

$$4\pi r^2 \Delta G_v = 8\pi r \gamma$$

$$r = \frac{2\gamma}{\Delta G_v} \tag{5.5}$$

We call this value the *critical radius* and give it the symbol r^*. By substituting r^* back into equation 5.4, the activation barrier for nucleation (ΔG^*) is found to be

$$\Delta G^* = \frac{16\pi \gamma^3}{3\Delta G_v^2} \tag{5.6}$$

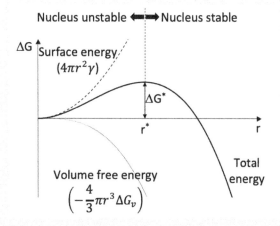

Fig. 5.5: Free energy as a function of particle size.

The value of ΔG_v is proportional to the latent heat of the material (how much heat is given out when it solidifies) and how far the liquid has been cooled below the melting point—the *undercooling*. The bigger the latent heat and the undercooling, the larger ΔG_v will be. This translates to a smaller critical radius (r^*) and a smaller energy barrier for nucleation (ΔG^*), indicating that the nucleation process is easier. In practice, the barrier to nucleation of a spherical particle within the bulk of a liquid is quite high, so nucleation tends to happen at surfaces such as the solid wall of a mould because this effectively reduces the energy barrier for nucleation and speeds up the process.

5.3   Mechanical properties of metals

Niobium-titanium is the workhorse superconductor, not because it has amazing superconducting properties, but because it has much better mechanical properties than the other technological superconductors. So it is worth looking at what gives metals their excellent mechanical properties and allows them to be easily made into wires. When a rod of metal is stretched by pulling the ends in opposite directions, the initial response is elastic. This means that if we stop pulling on it—remove the tensile force we are applying—it will return to its original length. What is actually happening is that the bonds between the metal atoms are being stretched. The stronger the bonds, the more force we need to stretch them by a certain amount, and the stiffer the metal rod will be. In this elastic regime, if the force is removed the metal rod will go back to its original length. It essentially behaves in the same way as a spring and follows Hooke's law: the extension of the rod (how much longer it gets) is proportional to the force that is applied. The constant of proportionality is called the spring constant and it relates to how stiff the sample (or spring) is. We can write Hooke's law as $F = kx$ where F is the force, x is the extension and k is the spring constant.

Ideally, materials scientists like to define things like stiffness as *materials properties*—quantities that do not change with the dimensions of the object that is being tested. An example of a materials property that we have already encountered is *resistivity* which depends on the material and temperature but (unlike resistance) is not influenced by the size and shape of the wire. Intuitively we know that the thicker the diameter of a metal rod, the more force is needed to extend it by the same amount. Essentially this is because the force we apply gets distributed over more bonds and so the force we are applying per bond has decreased. If you model the original rod as a single spring, doubling the cross-sectional area of the rod is like putting another identical spring in parallel with the first and pulling on both together, as shown in Fig. 5.6(a). The force pulling each spring is half of the original amount. Therefore, it

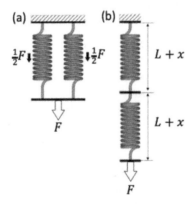

Fig. 5.6: Spring model for the elastic properties of a material. (a) Two springs connected in parallel are equivalent to doubling the cross-sectional area of the sample. (b) Two springs connected in series are equivalent to doubling the sample length.

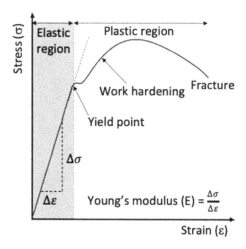

Fig. 5.7: Typical stress–strain curve of a ductile metallic material.

is useful to introduce a quantity called *stress* (σ), which is defined as the force per unit cross-sectional area (A) of the sample: $\sigma = \frac{F}{A}$. We also have to consider the extension term in a similar way. If we double the length of the rod, we would expect the extension to double. This is because each bond along the length of the rod is being stretched by the same amount. The longer the rod, the more bonds there are along its length and so the greater the total extension will be. This is simply like connecting two identical springs end-to-end in series (Fig. 5.6(b)). When we apply a force, F, we would expect each spring to extend by a distance x and so the total extension will be $2x$. So, instead of using the total extension of the rod, we use a quantity called *strain* (ε) which expresses the change in length of the rod x as a fraction of its original length L: $\varepsilon = \frac{x}{L}$.[5] We can then rewrite Hooke's law in terms of stress and strain: $\sigma = E\varepsilon$, Here, we have defined a new quantity E, which we call the *Young's modulus*. Just like resistivity, it is a materials property that does not depend on the dimensions of the sample and is a measure of the material's stiffness.[6] Figure 5.7 shows a typical stress–strain curve for a metal. The elastic region is where Hooke's law is obeyed and the stress increases linearly with strain. The gradient of the stress–strain curve in the elastic region gives the Young's modulus of the material.

So what happens if we keep pulling on the metal rod? Eventually the material will *yield* and something called *plastic deformation* starts to occur. When this happens it gets easier to extend the rod and the stress–strain curve flattens out. Inside the material, instead of the bonds just carrying on stretching, a different mechanism called *slip* is activated. Crystal defects called *dislocations* start to move under the influence of the applied stress. Applying a tensile force along the rod produces a shear (sliding) force

---

[5] Formally this definition of strain is known as the *engineering strain*.

[6] It is simple to show that the spring constant, k, is related to Young's modulus, E, by the equation $k = \frac{EA}{L}$.

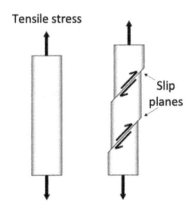

Fig. 5.8: Slip along planes tilted at 45° to the applied tensile stress.

along the diagonal direction (45° to the axis of the rod), as shown in Fig. 5.8. This has the effect of the diagonal crystal planes trying to slide over one another. If the crystal was perfect (no dislocation defects) it would take an enormous shear stress to shift the planes over one another in one go, but dislocations provide a much easier way for this shear to take place. A good way of thinking about how they work is by taking the example of trying to move large carpet across the floor. If we just hold it at one edge and give it a tug, we need to pull quite hard to get it to move because of friction. An easier way would be to make a ruck at one edge of the carpet and then push it across the carpet to the other side. The force we have to apply to move the ruck is much lower than the force we would need to move the whole carpet in one go, but we would only have succeeded in moving it by a short distance. We would need to repeat the process over and over again, moving the carpet a little bit sideways each time.

This is exactly what happens with dislocations. They are linear crystal defects—like the ruck in the carpet—and each time one moves across from one side of the sample to the other, the crystal planes shift by a small distance. The offset produced by each dislocation is called the *Burgers vector* and it depends on what kind of crystal structure it has and what kind of dislocation it is, but it is of the order of the spacing between atoms. Millions of dislocations need to move through the material to get appreciable extensions to our sample, so if we looked inside a material that has been plastically deformed, using a powerful electron microscope, we would see a high density of these linear defects. Interestingly, once we have enough shear stress to get the dislocations to start moving, they will carry on moving if we keep the stress roughly constant, which is why the graph flattens off. If we remove the applied stress, the dislocations do not move back, so the extension produced by slip is permanent (unlike the reversible behaviour of the elastic region). The sample is said to have undergone plastic deformation. Materials in which it is easy for dislocations to move are said to be *ductile*, whereas those in which dislocation motion is difficult will be *brittle*. Metals tend to be ductile because metallic bonding is not directional and it is relatively easy for the atoms to move their positions slightly as a dislocation passes through. In contrast, it is much more difficult to generate and move dislocations in ceramic materials with ionic or covalent bonding.

Yield—the onset of plastic deformation—is not the end of the story. When we have millions of dislocations all moving through the crystal on different slip planes, they end up getting tangled up with each other. This stops them from moving and there becomes a backlog of dislocations piling up. To free them up and enable them to move again, we typically need to increase the stress that we apply. This is because each dislocation will have a preferred crystal plane that it travels along, and we need to get it to switch to another (less favourable) plane to get past the obstacle. The material therefore gets harder—the gradient of the stress–strain curve increases again. We call this *work-hardening* because it happens as a result of mechanically 'working' the material. Eventually the dislocations are so tangled up inside the material that fracture occurs instead (the bonds break).

5.4 Thermomechanical processing of NbTi

Now we will turn our attention to how we optimise the critical current density in a niobium-titanium alloy. We know from the discussion in Section 4.1 that in type II superconductors like NbTi we need to pack the material full of structural defects to be able to get high currents without generating resistance. In fact, if we had perfect single phase NbTi, it would carry almost no resistance-less current because there would be nothing to stop flux lines moving and dissipating energy. It turns out that there is a really neat way of doing this in NbTi, and the phase diagram gives us the clue. We know that one of the ways to introduce pinning centres in superconductor is to introduce non-superconducting secondary phase particles. As luck would have it, a β NbTi alloy can actually lower its free energy by forming some particles of non-superconducting α Ti. By controlling this precipitation process we can obtain fantastic flux pinning without having to add in anything else. In principle this can be done by taking the supersaturated β alloy (the one with too much Ti dissolved in it) and heating it up a bit so that the atoms can move (diffuse) fast enough for the α phase to nucleate and grow. By tweaking the exact temperature and length of time of this heat treatment (called an *anneal*) we can control the number and size of the α precipitates to optimise pinning.

In practice, in the NbTi system it is actually quite difficult to persuade the α phase to precipitate even though it is thermodynamically favourable for it to do so. One effective way we can speed up the process is by putting in some mechanical deformation. Because the work-hardened material contains a high density of dislocations which raise its free energy, there is a big driving force for it to regrow as defect-free material if we heat it up enough to allow atoms to diffuse. This process is known as *recrystallisation* because new crystals nucleate and grow to consume the defective material. As this recrystallisation process happens, α phase precipitation is facilitated, so by mechanically working (physically deforming) and then annealing (heating up) the alloy we can increase the fraction of non-superconducting α phase.

This really plays into our hands because we want to manufacture thin wires of NbTi. This involves starting with a large billet of the material (a big cylindrical piece) and

The Wider View

Precipitation hardening

Solid state precipitation is used in structural alloys known as precipitation hardened alloys. The archetypal alloy of this type is Al–4 wt% Cu. Aluminium is a useful metal for automotive and aerospace applications because it has a low density, good ductility and good corrosion resistance. However, pure Al is far too soft to be useful. By adding a few weight percent of Cu and carrying out a heat treatment similar to the one described for NbTi, we can form secondary phase precipitates that obstruct the passage of dislocations. Because dislocations motion is impeded, the material work-hardens more effectively, enhancing its mechanical strength. In this case, we need to control the heat treatment to optimise the size and spacing of the precipitates, as summarised in Fig. 5.9(a). In the early stages of the precipitation anneal, metastable phases form first. These are not actually the thermodynamically most stable phases but they form very quickly because their crystal lattice closely matches the host crystal—they are said to be *coherent* and they have a much lower interface energy than incoherent precipitates. However these fine, coherent precipitates can easily be sliced through by dislocations because the crystal structure is not very distorted and so they are not very good at strengthening the material. If, on the other hand, we let the precipitates grow too big, they become less coherent with the host lattice. That makes it more difficult for dislocations to cut through them. Unfortunately, as they grow, the precipitates get further apart and the dislocations can then bow around them instead (Fig. 5.9(b)). For this reason we need to strike a happy medium to optimise the hardness of the alloy.

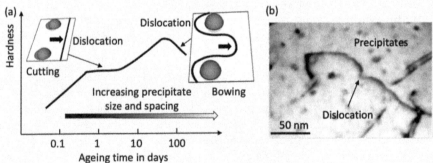

Fig. 5.9: (a) Optimising strength of Al-Cu alloy by ageing. (b) TEM of dislocations in steel bowing around copper precipitates (courtesy of S. Lozano-Perez, University of Oxford).

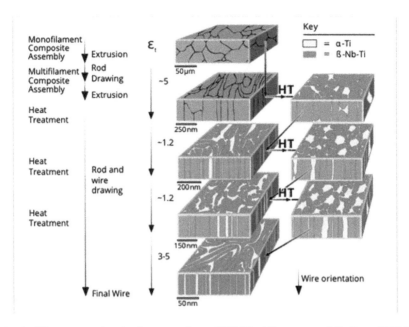

Fig. 5.10: Thermomechanical processing of NbTi. (Courtesy of P. Lee, NHMFL.)

extruding it or drawing it through a series of increasingly narrow nozzles called dies. Extrusion is a process just like pushing toothpaste out of a tube, and it is typically used in the early stages of the process. Wire-drawing is used later in the process and involves grabbing hold of the end of the wire and pulling it through the die under tension. To get a thin wire we cannot just keep pulling it through narrower and narrower dies because the metal would become so work-hardened that it would break. Instead, we draw it down a bit, putting in some mechanical damage, and then we do a heat treatment to recover the ductility and encourage some α phase to precipitate. Then we draw it some more, and heat it again. This sequential process is repeated until the desired final wire diameter is obtained. Figure 5.10 shows a schematic diagram of the microstructure that is obtained at different stages of the process. As you can see, not only do you end up increasing the total volume of α phase, but the process also has the effect of decreasing the size of the α precipitates by a factor of about 1000. (Note that the scale bar starts off at 50 micrometres and finishes off at 50 nanometres.) The final precipitate size can be controlled by selecting the original billet dimensions, final wire diameter and heat treatment parameters to end up with the desired microstructure. This is what materials science is all about—using our understanding of materials systems to design materials (and strategies for processing them) to optimise their microstructure and hence their performance.

5.5 Large Hadron Collider wires

So far, we have talked about how we can engineer a two phase microstructure in NbTi to optimise flux pinning by exploiting the thermomechanical process that we use for making thin wires. But for real applications, we cannot use single strands of NbTi wire. We actually need many fine filaments of superconductor, all embedded in plenty of copper to keep the currents (and magnetic fields they produce) as stable as possible. Mechanical vibrations (for example) can lead to flux lines jumping from one pinning site to another and dissipating some heat. If we cannot get that heat out of the place where it has been generated, an avalanche effect happens. Hot spots generated by these flux jumps create small regions that stop superconducting so that the supercurrent has to divert around them. Effectively the cross-sectional area of superconductor for the current to flow through is reduced so the current gets concentrated in the vicinity of a hot spot. If the current locally exceeds the critical current density, the rest of the wire will lose superconductivity and a huge amount of energy suddenly has to go somewhere. The copper surrounding the filaments of superconductor is essential for extracting the heat generated and stopping these quenches. The copper also produces a good electrical shunt for current to pass through if the superconductor does become resistive. In the magnets used to bend and focus the proton beam in the Large Hadron Collider (LHC), the 1 mm diameter wire strands contain no fewer than 6300 separate filaments of NbTi (~ 0.006 mm in diameter), each surrounded by a layer of copper as shown in Fig. 5.11.

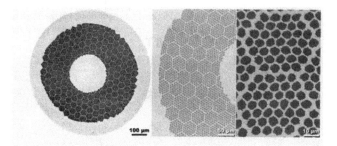

Fig. 5.11: LHC wire cross section. Reproduced from (Rogalla and Kes, 2012) with permission of Taylor and Francis Group, LLC, a division of Informa plc.

So how on earth do we go about making wires many kilometres long with thousands of really tiny but continuous filaments all embedded in copper? Luckily, copper and NbTi have rather similar mechanical properties. To make the wire, the starting material is a cylindrical billet made by assembling a rod of NbTi inside a Cu tube, sometimes with an extra thin layer of Nb in between. When this composite billet is extruded, the separate components extrude together, so the final object is just a narrower diameter version of the original—NbTi surrounded by a layer of Nb and then a layer of Cu (Fig. 5.12). This co-extrusion method is also used for getting stripy toothpaste and it is how sweets like seaside rock are made, with letters or patterns that stretch throughout the whole length. After initial extrusion that reduces its diameter, the superconducting

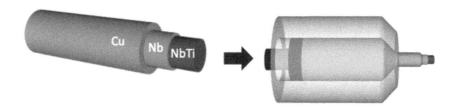

Fig. 5.12: Initial stages of processing NbTi wire.

wire is passed through a hexagonal die. The resulting hexagonal cross-section rods can then be packed together in a honeycomb arrangement, placed inside another copper tube, and the extrusion process can be repeated to create a wire with a bundle of filaments neatly arranged inside. These multifilamentary wires are then again passed through a hexagonal die, stacked together in the same way as before, and extruded again to produce a wire with a hierarchy of hexagonal structures inside. This is then drawn down to the final diameter required (Fig. 5.13).

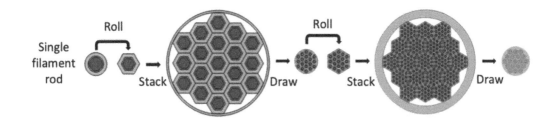

Fig. 5.13: Multifilamentary wire processing method.

This still is not the end of the story. We also often need to twist each strand and weave multiple strands together to make a cable. This is essentially to minimise the electromagnetic cross-talk between the filaments that can introduce unwanted losses. Rutherford cable is a particular type of cable that is widely used in accelerator magnets, where the magnetic field quality needs to be excellent. The Rutherford cable in the Large Hadron Collider magnets (shown in Fig. 5.14) contains 36 strands, and each strand contains 6300 NbTi filaments. In total the LHC uses 7500 km of Rutherford cable, which translates to over 250,000 km of strand (wire). If we lined up all of the individual NbTi filaments, they would stretch for about 1.5 billion kilometres—10 times the distance from the earth to the sun!

Fig. 5.14: Rutherford cable. (Courtesy of CERN).

Chapter summary

- NbTi is called the workhorse superconductor because it is by far the most commonly used superconducting material in large scale applications, but the materials science that has gone into optimising the wires is fascinating. Not only does it need to be made in complex multifilamentary wire geometries to get the required stability, but every strand has to be engineered on the nanoscale to optimise its current carrying capacity at the operating fields.

- The precise geometry and architecture of the wire—number and diameter of filaments, how much copper is used, size and distribution of the α Ti precipitates—is tailored to the specific application.

- Large Hadron Collider wires contain far more filaments than the wires used for MRI magnets and their microstructures are optimised for operation at lower temperatures and higher magnetic fields.

- The Nb-Ti phase diagram has been examined in some detail because phase diagrams are key tools that materials scientists use for designing alloys and processing strategies. The idea of controlling precipitation of secondary phases in alloys is a major theme in materials science.

6

Quirky Quantum Effects

By the end of the 1800s, physicists had figured out a set of rules that do a really good job of describing how the world around us works. Collectively, these theories are known as *classical physics*. The way in which physical objects like snooker balls move could be explained and predicted accurately using Newton's laws of motion, and the behaviour of light could be explained using Maxwell's theories of electromagnetism. In classical physics, objects (particles) have well-defined properties like mass and charge and their behaviour is described by quantities such as kinetic energy and momentum which depend on the mass of the particle and how fast it is moving. Light, on the other hand, behaves as a wave which is essentially a vibration—a disturbance—that carries energy from place to place, but does not transfer any material. There are lots of different kinds of waves—water waves, sound waves, waves on strings—which share some common properties even though the nature of the disturbance is different in each case. Water waves are essentially local peaks and troughs in the height of the water, whereas waves are produced in stretched strings by pulling the string sideways and then letting go. Think of a double bass player plucking the strings or a violinist drawing their bow across the string. We hear sound from these instruments because the strings make the wooden front of the instrument vibrate, which sets up pressure waves (sound waves) in the surrounding air. Light is a special kind of wave called an electromagnetic wave. They consist of regular oscillations of electric and magnetic fields and they travel through space at a constant speed—the speed of light. How bright the light is depends on how much the electric (and magnetic) field changes by during the oscillation—the *amplitude* of the wave. On the other hand, colour depends on the *wavelength* of the light, which is the distance in space that is taken up by a single oscillation, with red light having a longer wavelength than blue light (Fig. 6.1). In fact, there is a huge spectrum of wavelengths of electromagnetic waves spanning from about 100,000 kilometre (10^8 m) in 'extremely low frequency' radio waves to 1 picometre (10^{-12} m) in gamma rays—20 orders of magnitude! Visible light occupies a tiny part of this spectrum–about 400-700 nanometres (4–7×10^{-7} m). Classical wave physics describes many of the phenomena of light that you will already be familiar with, like the bending of light (refraction) when it enters a material, reflection of light from a shiny surface and diffraction (spreading out of light when it passes through a small hole).

However, despite the huge success of classical physics, there were some experimental observations that could not be explained using this set of rules. One of these was the

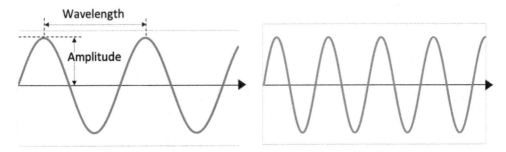

Fig. 6.1: Red and blue light waves.

fact that if you take a tube of hydrogen gas at a low pressure and apply a very high voltage across the ends, a pink glow is produced that is made up of light of a series of specific colours (wavelengths). This is known as the hydrogen emission spectrum. Classical physics could not explain why there are special, distinct wavelengths of light emitted, rather than a continuous spectrum of all the different colours. Another puzzling problem related to something termed *black body radiation*. A black body is a theoretical object that perfectly absorbs all of the radiation (light) that falls onto it and then re-emits it. Lord Rayleigh (famed for explaining why the sky is blue) predicted using classical physics that the intensity of the light emitted would increase drastically as the wavelength of light decreases, resulting in an infinite amount of energy being emitted at ultraviolet wavelengths (which are slightly shorter than visible violet light). This 'ultraviolet catastrophe', as it is called, is obviously unphysical and would make life on Earth impossible! The breakthrough to understanding the experimental observations came in 1900 when Max Planck made the inspired hypothesis that instead of energy being continuous, it comes in discrete packets known as *quanta*. Importantly, he realised that the energy of a single quantum of light, which we now call a *photon*, is inversely proportional to its wavelength.[1] A photon of red light has a longer wavelength and therefore a lower energy than a photon of blue light.

At the same time as scientists like Planck were puzzling over these problems that could not be solved with the classical theories of light that they had at their disposal, some giant leaps were also being made on understanding the structure of atoms, triggered in 1897 when J.J. Thomson discovered the existence of electrons. Typically we think of these as particles, and so it is natural to want to use Newton's laws to explain their motion. However, it was soon discovered that classical physics was not going to be able to adequately explain how electrons and atoms behave. Led by Nils Bohr, new theories of the atom were developed that took on board Planck's new ideas of quantisation of energy to explain concepts such as electrons occupying 'orbitals' with discrete, fixed energy levels. This explained things like the hydrogen emission spectrum because specific, discrete energies of light are emitted when electrons move from a high

---

[1]Planck's famous equation is usually written as $E = hf$, where E is the energy of the photon and f is the frequency of the wave—the number of vibrations per second—and h is a universal constant called Planck's constant. Since all frequencies of light travel at the same speed in vacuum, frequency is related to wavelength (λ) by the equation $f = \frac{c}{\lambda}$ where c is the speed of light.

energy orbital into a lower energy orbital of the hydrogen atom. But instead of the situation becoming clearer, the more the physicists tried to understand what was going on, the more confusing it all became. For example, in some experiments, electrons seem to behave like particles and in other experiments they behave more like waves. Similarly, light sometimes seems to behave like a wave, and in other circumstances it has to be treated as particles (photons) to explain the observations. These are two sides of the same coin. Essentially the problem came down to the fundamental physics of small particles like electrons not fitting into the binary, classical picture of things being either particles or waves. A whole new physics—quantum mechanics—had to be developed to describe their weird behaviour. One of the key fundamental ideas that comes out of this is that it is impossible to know everything about these particles. With a big object like a ball, we can simultaneously know where it is and what speed it is moving at. However, with an electron, if we try to pin down precisely where it is in space, we lose information about how it is moving. The more accurately we know its position, the less accurately we know its momentum. This is one manifestation of a more general principle called *Heisenberg's uncertainty principle*. We can only actually know how likely it is that the electron is at a particular place at a particular point in time, which is why in the modern picture of an atom we think of the electrons occupying hazy orbitals often described as 'clouds'. It is possible to describe a particle as a wave using a mathematical function called a wave function, where the square of the amplitude of the wave function tells us how likely the particle is to be at that point in space at that point in time.[2]

So why does this matter for understanding physical properties of materials, like electrical conductivity? Although the material itself is usually a large object, electrical conductivity results from the motion of electrons inside the material. Therefore, to understand electrical conductivity, and ultimately superconductivity, we need to understand the properties of the electrons in the material, and for that we need quantum mechanics.

6.1   Electrons in metals

Before we start thinking about what happens to the electrons in a superconductor, it is a good idea to remind ourselves about electrons in metals. The description of metallic bonding that you will probably be familiar with is a regular arrangement of positive ions in a 'sea' of electrons. The electrons essentially act as the glue that holds the positively charged ions together. Because they are *delocalised*—not tied to particular atom locations—they can easily move under the influence of an applied potential difference (voltage) generating a current (flow of charge). Not every electron in a metal is delocalised. Most of them stay in the electron shells around the atom, but the outermost (highest energy ones) are not held very tightly into the nucleus and the structure is stabilised by allowing these outer electrons to leave their atomic positions and spread out throughout the material. These delocalised electrons are responsible for the electrical conductivity of metals (along with a whole host of other properties), and

---

[2]In general, a wave function is complex (not real) and it is modulus squared of the wave function $|\psi|^2 = \psi^*\psi$ that gives the probability.

so they are sometimes called *conduction electrons*. We often use something called *free electron theory* to describe how the electrons in metals behave. Essentially we assume that the motion of each delocalised electron is independent of the other electrons and the positive ion cores and so they do not exert any forces on each other. This means that their energy is purely kinetic—from their motion. In real metals, the positive ion cores do actually produce a potential energy that influences the motion of the electrons slightly, but it is a weak interaction and we ignore it in free electron theory.

The next thing to consider is what energies these free electrons in metals have. This is a difficult problem though, because a chunk of metal, like a piece of wire, contains millions and millions of atoms, each providing at least one free (delocalised) electron. To simplify things, we will start off by thinking about a single isolated atom instead. You may be familiar with the idea that in an atom the electrons are arranged in shells around the nucleus. Each electron shell has a different energy, with the innermost shells (those closest to the nucleus) having a lower energy than the outer shells. Electrons confined to atoms can essentially only take the specific, discrete energy values associated with the particular shell they are in. This is an entirely 'quantum' idea that cannot be explained using classical physics. If we imagine filling our atom up with electrons, one by one, the first one would go into the first, lowest energy shell. The next electron would join it there, but then it is full. If we add another electron, it has to go into the second shell. This process continues until you have added all of the electrons.

So why can they not all sit in the lowest energy level? Surely that would mean the total energy would be lower? Yes, that is true, but electrons have a special quantum property that means that you cannot have more than one of them in exactly the same quantum state.[3] This is a very famous fundamental rule that particles like electrons have to obey called *Pauli's exclusion principle*. Two electrons are allowed in the first shell because there are two different flavours of electron—spin up and spin down—which means that they are not identical from a quantum mechanical point of view.[4] How about the second electron shell? You may have been taught that the second shell can actually hold eight electrons, which seems to contradict Pauli's exclusion principle. Well actually the second shell consists of four different *subshells* ($2s$, $2p_x$, $2p_y$ and $2p_z$) whereas the first shell only has one ($1s$). The 's' subshells are spherical in shape, but the 'p' subshells are shaped like dumbbells and can point in the x, y or z directions. Because of the effect of having multiple electrons in the atom that can screen some of the nuclear charge, it turns out that the energy of the 2s subshell is slightly lower than the energy of the 2p subshells, so electrons choose to fill the 2s subshell first, followed by the 2p subshells. Each of the subshells can hold a spin up and a spin down electron, so there are a total of eight electrons in the second shell (Fig. 6.2). The important thing is that electrons in each of these different subshells have a certain fixed energy

---

[3] A quantum state is a mathematical function that encodes all of the physical information about that particle such as where it is most likely to be in space, what its angular momentum is likely to be etc.

[4] This is the electron equivalent of the nuclear spin that was introduced in Section 2.2 when we discussed nuclear magnetic resonance.

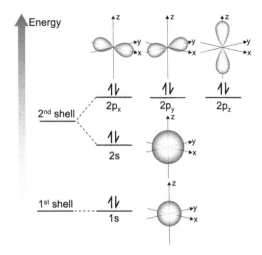

Fig. 6.2: Energy levels in an isolated atom.

value. If we put in the correct amount of energy we can promote an electron from a lower energy level to a higher energy level. The atom becomes unstable (because it has a higher energy than it did before), so an electron is likely to fall back into the hole left behind in the lower energy shell, releasing the excess energy, perhaps as a photon of light.

So what happens to these discrete electron energy levels when we bring lots of identical atoms (say copper atoms) together to make a chunk of metal? Let's think first about what happens when we just have two atoms close together. It is still the case that no two electrons can be in exactly the same quantum state. This means that each of the discrete states in the isolated atoms has to split into two states with different energies when we bring two atoms close enough together that their orbitals overlap (Fig. 6.3). The states necessarily get closer together in energy. Each time we add another atom, we have to split our energy levels again and the energy spacing between the levels gets smaller still. This means that in a chunk of copper containing a large number of atoms, this splitting of the states has happened so many times that the difference in energy between the states is incredibly small. Now, to understand the electronic properties, we can essentially treat our piece of copper as a large box into which we add electrons one by one. The first electron will occupy the lowest energy state. However, Pauli's exclusion principle still holds, so as more electrons are added, they have to go into higher energy states, just like they had to in an isolated atom. The difference is that these states are very close together in energy in a large piece of metal compared to the energy levels in isolated atoms.

The fact that the energy levels are so close together in macroscopic metal objects that the energy is effectively continuous means that classical physics can do a surprisingly good job at describing materials properties. One example is the classical Drude model for electrical conduction in metals which treats the delocalised electrons just like gas

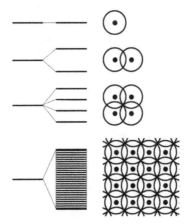

Fig. 6.3: Repeated splitting of energy levels when atoms are brought close together.

molecules. If we take a length of copper wire, for example, and connect it up to a power supply, the voltage produces a force that accelerates mobile electrons along the wire generating a current. If there were no obstacles in the way, Newton's law tells us that if the power supply was turned off the electrons would keep going at a constant velocity. However, real metals contain a forest of positive ion cores that get in the way of the electrons. When electrons collide with an ion core they are scattered in a random direction, so their average velocity along the wire will be lost (Fig. 6.4). This is the origin of electrical resistance and it means that if the power supply is switched off, the current will very quickly stop flowing. The kinetic energy of the electrons is transferred to the ions during repeated collisions thus generating heat. This simple, classical model reproduces some of the key properties of electrical conductivity in metals, but often the numerical values that are predicted do not match experimental values. The reason is that the model treats electrons as classical particles and fails to take into account the consequences of Pauli's exclusion principle.

As we have seen, free electrons in a metal fill up the energy states from the lowest energy upwards because no two electrons can occupy the same quantum state. The energy of the very last electron that we put into our metal 'box' is called the *Fermi*

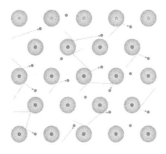

Fig. 6.4: Electron scattering by ion cores in a conventional metal.

energy. At absolute zero, all of the quantum states with energy below the Fermi energy will be occupied with an electron, and all of the states above the Fermi energy will be empty. However, as we increase temperature, the thermal energy smears out the boundary somewhat because electrons can be promoted from states slightly below the Fermi energy to those slightly above it. Although the electrons are all whizzing about, if we are not applying a voltage, they are equally likely to move in all directions, so there will be no net current flowing along the wire. When we apply a voltage, there is a force exerted on the electrons that will try to accelerate the electrons (as in the simple Drude picture). The difference is that most of the free electrons in the metal cannot actually respond. Take, for example, an electron that is initially stationary. It cannot start moving because there are no suitable empty states with slightly higher kinetic energy for it to move into. They already have electrons in them and Pauli's exclusion principle tells us electrons cannot share the same state. It is therefore only actually possible for the highest energy electrons to move—the ones near the Fermi energy where there are some empty states to move into. The upshot is that not all of the electrons will contribute to the current, so Sommerfeld modified Drude's original model to take the electrons' quantumness into account.

6.2 Electrons in superconductors

Having got the basic idea of the electronic structure of conventional metals, the time has come to find out about why there is no resistance in a superconductor—the question that you have probably been asking yourself right from the start of this book! If we consider the picture of a metal consisting of positive ion cores surrounded by a sea of delocalised electrons, it would be natural to think that the electrons would always repel one another because they all have negative electric charge. But there is a strange thing that can happen. Sometimes it is possible for there to actually be a very small attractive force between two electrons and they pair up to form something called a 'Cooper pair'. It is these Cooper pairs that carry the resistance-free current in a superconductor. To fully understand the physics of the superconducting state we cannot just consider isolated Cooper pairs moving through the material, but it is a good starting point.

First of all, let's think about how it might be possible for there to be an attractive force between two electrons. The answer is that, in real materials the electrons not only interact with one another, but they also interact with the lattice of ion cores. Using a classical picture of what is going on, if an electron moves, it will try to pull

Fig. 6.5: Visualisation of the Cooper pair formation mechanism.

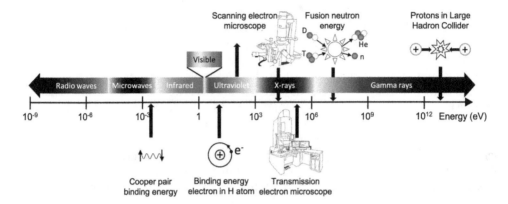

Fig. 6.6: Energy scale showing how the Cooper pair binding energy relates to the energy of single photons of electromagnetic radiation as well as various other particles.

the neighbouring positive ions with it because of the electrical attraction between them. The ions are arranged in a lattice (regular arrangement) and behave a bit like an array of masses connected by springs. The lattice distorts in the vicinity of the electron, increasing the concentration of positive charge around the position of the electron, as shown in Fig. 6.5. However, the electron is very light and can move quickly whereas the ion cores are much heavier and respond much more slowly. So, by the time they have concentrated around where the electron was, it has moved away. The excess positive charge left behind in its wake can attract a second electron. The consequence is that effectively there is an attractive force between the pair of electrons involved, via the distortion of the lattice of positive ions. There are lots of issues with this simple visualisation of what is going on, but it gives the essence of what is going on. Quantum mechanically speaking, the vibrations of the positive ion lattice are quantised in energy and can be thought of as particles called *phonons* with a fixed amount of energy—not to be confused with *photons* which are particles of light. The energy saved by two electrons pairing up to form a Cooper pair is absolutely miniscule—just a couple of thousandths of an electronvolt ($\sim 10^{-22}$ joules), where an electronvolt is the amount of kinetic energy an electron would have if it had been accelerated through a potential difference of 1 volt. To set that in context, Fig. 6.6 shows typical energies associated with particles in various situations. As you can see, the Cooper pair binding energy is about 10,000 times smaller than the energy required to remove an electron from a hydrogen ion to ionise it, and about a million times smaller than the energy carried by a single X-ray photon! It is pretty remarkable that such a small energy saving can be responsible for such a dramatic phenomenon as superconductivity.

So the question now is, 'How come these Cooper pairs can move without any resistance?' To answer this, we need to go back to what we know about normal, unpaired electrons. As already discussed, normal electrons in metals have to obey Pauli's exclusion principle. This is because they have the quantum property called spin. However, Cooper pairs consist of two electrons, one spin up and one spin down, so their overall spin is zero. This means that they do not have to obey Pauli's exclusion principle any longer. We won't go into the details of why that is the case—you can read about that in books on quantum physics—but the upshot is that all of the Cooper pairs can occupy the *same* quantum state. They are all allowed to have the same energy and momentum and because the system will want to lower its energy as much as possible, they will all go into the lowest energy state. We call this process *condensation* which is why in Section 3.5 we referred to the energy saved by the material becoming superconducting as the *condensation energy*.[5] The effect of this is that there is a small energy gap between the paired electron states (lower energy) and the unpaired energy states (higher energy). If we start a current moving along the superconducting wire, all of the Cooper pairs will move with the same momentum (same velocity) as each other because they have condensed into the same low energy quantum state. Because of the energy gap, it is much easier for all of the Cooper pairs to drift along with the crowd at the same speed as all of the others, than to try to do their own thing. This means that a Cooper pair cannot be scattered by an obstacle into a quantum state with a different momentum. This is, in essence, why they can move with zero resistance—they do not get deflected from their path in the same way that normal electrons in a metal do. That is not to say that they do not get scattered at all. In fact they interact with the lattice all the time. It is, after all, the thing that makes them form Cooper pairs in the first place. The key difference is that they can only be scattered into states that leave the pair of electrons with the same overall momentum. A colleague in Oxford, Professor Stephen Blundell, describes the superconducting state as being like the Terminator from the famous films because nothing can stop it—it just keeps on going relentlessly (Blundell, 2009). Alternatively, you can compare the motion of Cooper pairs in a superconductor with an army marching in unison. Unpaired electrons in a normal metal behave more like a crowd of people milling about.

---

[5]The word *condensation* comes from the Latin for *making more dense*. This is why we use it to describe the gas to liquid transition (the liquid is denser than the gas). In the case of a superconductor the term arises because the 'density' refers to how many electrons there are within a certain energy range, and because they all collect together in the same energy state, we call this a condensate and we say the electrons have condensed into the superconducting state.

Under the Lens

Superconducting energy gap

In Chapter 3, the concept of *condensation energy* was introduced as the amount of energy saved by the material when it becomes superconducting when no magnetic field is applied. Using thermodynamic arguments we found that the condensation energy scales with the square of the critical field of the superconductor.

$$\text{Condensation energy} = \frac{VB_c^2}{2\mu_0} \tag{6.1}$$

where V is the volume of the superconductor and $\mu_0 = 4\pi \times 10^{-7}$ H m^{-1} is the permeability of free space. If we take niobium (Nb) as an example, it has a critical field of about 0.2 T at 0 K. This means that its condensation energy per unit volume is about 1.6×10^4 J m^{-3}. The superconducting gap (2Δ), shown in Fig. 6.7, is simply the condensation energy per Cooper pair, so to estimate it from the condensation energy we need some idea of how many Cooper pairs there are in 1 m^3. The easiest way of doing this is using information on the crystal structure of Nb (although we could use density and mass if preferred).

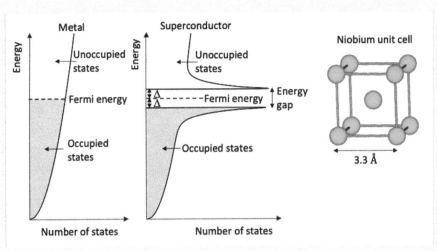

Fig. 6.7: Diagrams showing the energy states in a metal and a superconductor along with the BCC unit cell of niobium.

Niobium has a *body centred cubic* (BCC) unit cell with a lattice parameter—side length of the cubic unit cell—of 3.3 Å = 3.3×10^{-10} m, as shown in Fig. 6.7. This means that the volume of the unit cell (Ω) is $(3.3 \text{ Å})^3 = 3.6 \times 10^{-29}$ m^3. If you look carefully at the BCC unit cell, only $\frac{1}{8}$th of each of the corner atoms is within this unit cell (with the rest being in neighbouring cells), so the eight corner atoms only contribute a total of one atom to each unit cell. Adding this to the atom at the body centre of the cube gives a total of two atoms per unit cell in BCC structures. Using this information we can find the volume per atom and the condensation energy per atom.

$$\text{Volume per atom} = \frac{\Omega}{2} = \frac{3.6 \times 10^{-29}}{2} \approx 1.8 \times 10^{-29} \, \text{m}^3 \tag{6.2}$$

$$\text{Condensation energy per atom} \approx 1.6 \times 10^4 \times 1.8 \times 10^{-29} \approx 3 \times 10^{-25} \, \text{J} \tag{6.3}$$

This looks like a minuscule amount of energy, but let's try to put it into context. The first thing to do is to change the units into something more convenient. Our favourite choice of energy scale is the *electronvolt* (eV). Now 1 eV is the electrical work done when a single electron is accelerated through a potential difference of 1 volt. Electrical work done is just given by the charge multiplied by the potential difference, so 1 eV = electronic charge $\times$ 1 V = 1.6×10^{-19} J. So the condensation energy per Nb atom works out to be about 2×10^{-6} eV.

Now if we want to get an idea of the superconducting energy gap—the energy binding a Cooper pair together—we need to estimate how many Cooper pairs there are per atom. Unfortunately there is no simple way to do this, but it turns out (if you go through the full microscopic theory of superconductivity) to be of the order of one Cooper pair per 1000 atoms. This is because it is only electrons close to the Fermi level that can pair up. This means that the energy saved per Cooper pair—the superconducting gap energy at 0 K ($2\Delta(0)$)—is about $2 \times 10^{-6} \times 1000 \approx 2 \times 10^{-3}$ eV ≈ 2 meV (millielectronvolts).

We can use the $2\Delta_0$ result to get an idea of the critical temperature of Nb. The amount of thermal energy (in joules) available at a particular temperature is $\sim k_B T$ where $k_B = 1.38 \times 10^{-23}$ J K^{-1} is the Boltzmann's constant. So if this thermal energy is used to break apart a Cooper pair we can get a rough estimate of the transition temperature.

$$2\Delta_0 \approx meV \approx 3 \times 10^{-22} \, \text{J} \tag{6.4}$$

$$k_B T_c \approx 3 \times 10^{-22} \tag{6.5}$$

$$T_c \approx 20 \, \text{K} \tag{6.6}$$

This turns out to be a bit of an overestimate of the real T_c of Nb which is actually about 9 K, but if we use a more accurate relationship from the full microscopic theory, $2\Delta = 3.5 k_B T_c$, we end up with a value of 6 K, which is a bit closer.

We are not going to dig any deeper into the theory of superconductivity here because this book is about materials science rather than quantum physics, but there are a couple of other things that are worth drawing to your attention at this point. The first is that Cooper pairs are not like two snooker balls glued together. They are much more like two people across a crowded room catching each other's eye. Depending on the material, they may be separated by up to about 100 nanometres, which is about 500 times bigger than the atomic spacing in the material. The fundamental property that tells us the separation of electrons in a Cooper pair is called the *coherence length*. In low temperature superconductors this can range from a few nanometres right the way up to 100 nm, so we would expect loads of Cooper pairs to exist within the same volume of material. High temperature superconductors are a bit different because the coherence length is much smaller (about 1 nm) so electrons in the Cooper pair are much closer together (Fig. 6.8). It turns out that this makes a big difference to how defects in the crystal lattice affect current transport. This is for two main reasons. Firstly, as discussed in Section 4.1, the radius of the normal core of flux lines in the material is of the order of the coherence length, so it will influence what kind of defects are the best flux pinning centres. Secondly, the coherence length affects another special quantum effect called *quantum tunnelling*.

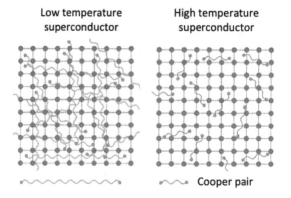

Fig. 6.8: Cooper pairs in typical low temperature and high temperature superconductors.

The other takeaway message is that not all of the outermost conduction electrons (the ones near the Fermi level) will pair up to form Cooper pairs. At real operating temperatures there will be enough thermal energy around to break up some Cooper pairs because their binding energy is so small. This means that there will be a mixture of paired electrons (in states below the energy gap) and unpaired electrons (in states above the energy gap) that are also free to move. Luckily, this does not lead to resistance under steady current conditions because all of the transport current is carried by the Cooper pairs. The normal electrons that are also present will essentially be shorted out. However, the same is not true if the current is changing, so superconductors do not have zero resistance under alternating current conditions.

6.3 Quantum tunnelling

Now that we have some idea of what the electrons are doing in a superconductor, it is interesting to consider what happens when the Cooper pairs meet a barrier—for example a thin region of non-superconducting material. There is a really funky phenomenon that happens when quantum particles meet some sort of wall that gets in the way of their movement. We call these obstacles *potential barriers* because they are regions where the potential energy of the particle has to increase. We can think of a potential barrier as a wall. If we roll a classical particle like a ball towards the wall, it will probably bounce back the way it came, but if the wall is low enough, the ball may have enough kinetic energy to get up and over it (Fig. 6.9a). The higher the wall, the bigger the potential energy barrier and so the more kinetic energy the ball must have to start with if it is to have any chance of making it to the other side. If the initial kinetic energy of the ball is less than the potential energy it has to gain to get over the wall, it will always bounce backwards. This is the classical mechanics of particles and is based on the principle of conservation of energy. Since energy cannot be created, the ball cannot raise its potential energy to a level higher than its original kinetic energy. Interestingly, small particles like electrons and Cooper pairs do not behave like classical particles. When they hit a potential barrier, there is a chance that they can get to the other side by tunnelling through the wall rather than going up and over it (Fig. 6.9b). This means that the electrons do not actually need to have more kinetic energy than the potential energy of the barrier to get to the other side. This phenomenon is called *quantum tunnelling* and there is no classical equivalent of it.

It may sound like the particle is breaking the principle of conservation of energy, but actually it is not. The difference between a classical particle (like the ball) and a quantum particle (like an electron) is that physical properties like energy do not, in general, take a single, well-defined value for a quantum particle. Instead, they exist in a condition that is a collection, or *superposition* of possible states, each with its own energy associated with it. You may have come across superposition in the context

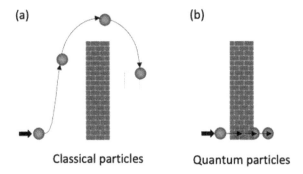

Fig. 6.9: (a) Classical particles going over a potential barrier, and (b) quantum particles tunnelling through a potential barrier.

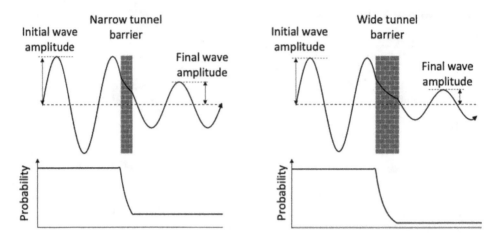

Fig. 6.10: Visualisation of waves associated with quantum particles tunnelling through a potential barrier.

of the interference of waves. It just means that we add them together. A quantum particle can be described by a superposition of waves, each with a different energy. The upshot of this is that, at any particular instant in time, the particle may be partly in the wall and partly outside it. The part that is inside the wall effectively has a negative kinetic energy, but the part outside will have a positive kinetic energy, and the average kinetic energy will always be positive. Luckily, it is only the average value that is related to the physical property of the particle itself, so we do not need to worry that the kinetic energy is negative in some regions of space. Phew! The chance that an electron manages to tunnel through the potential barrier depends on both the height of the barrier (analogous to the height of the wall) and also how wide it is (the thickness of the wall), as shown schematically in Fig. 6.10. This means that the tunnelling current—which is proportional to the number of electrons that tunnel through per second—also depends on the height and width of the potential barrier.

The Wider View

The scanning tunnelling microscope

Quantum tunnelling is an important phenomenon that we make use of in materials science in various ways, and it is no surprise from its name that the scanning tunnelling microscope (STM) is one of them. The STM is an amazing instrument that is capable of imaging individual atoms that make up the surface of a material (Fig. 6.11(a)). So how does this amazing microscope work? Basically you take a piece of metal wire (special stuff, not any old piece of wire) and electropolish it into a really sharp point with a radius at the tip as small as 50 nanometres. Inside a high vacuum chamber, a voltage is applied to the tip and is then brought very close to the surface of the sample (typically about a tenth of a nanometre (10^{-10} m). This is close enough that current can tunnel between the tip and the sample through the vacuum gap (potential barrier). To image the sample, the tip is either scanned backwards and forwards across the surface of the sample and the tunnelling current is recorded (constant height mode), or a feedback loop is used to keep the tunnelling current constant by moving the tip physically up and down to follow the contours of the sample's surface (constant current mode). In this latter mode, the distance that the tip has to move to keep the tunnelling current constant gives a measure of the height of the surface features (Fig. 6.11(b)). If we keep the tip still and change the voltage on the tip systematically, the tunnelling current then provides a measure of how many electronic states there are as a function of energy in the atom closest to the tip. This is a technique called *Scanning tunnelling spectroscopy* (STS) and we can use it to measure the superconducting energy gap in superconductors (Fig. 6.11(c)). Spectroscopy is a general term for measurements of the response of a material as a function of energy and we can do all sorts of different kind of spectroscopy using electrons, X-rays and infrared light depending on the energy scale we are trying to measure.

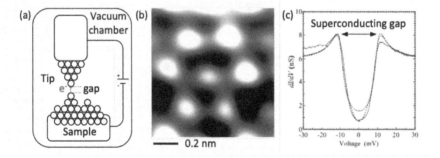

Fig. 6.11: (a) Diagram of an STM. (b) An STM image of the atoms in the surface of MgB_2. (c) STS results showing the superconducting energy gap in MgB_2. (b) and (c) are reproduced with permission from (Ekino *et al.*, 2007).

6.4   The weakest link

In superconductors, Cooper pairs can tunnel across narrow potential barriers that we call *weak links* or *Josephson junctions*.[6] This means that if we sandwich a thin slice of non-superconducting material between two slabs of superconducting 'bread', some of the Cooper pairs can tunnel from one side to the other. By tunnelling, we can get a certain amount of current through the non-superconducting layer without generating any resistance, and therefore without needing to apply any voltage. The maximum resistance-free current that can get across the junction depends on how thick the junction is, as well as things like whether the junction is a layer of insulator, normal conductor or just a less-good bit of superconductor. All we need to know at this point is that this junction critical current is considerably lower than the critical current density of the superconducting slabs on either side. This ability to get a certain amount of current across a junction without needing a voltage is known as the DC (direct current) Josephson effect and is illustrated in Fig. 6.12.[7] The magnetic properties of weak links are also worth mentioning. If we apply an external magnetic field to our sandwich of superconductor-barrier-superconductor in an external magnetic field, the critical current of the junction rapidly decreases to zero as the field increases and then rises and falls repeatedly forming a series of fringes that look just like the bright and dark interference fringes that are produced when light is passed through a slit (a Fraunhofer interference pattern). In fact, since you can think of the superconducting state as a wave, it is maybe not surprising that these devices exhibit wave-like properties. The Josephson effect may seem to be something that is only of academic interest, but actually Josephson junctions form the fundamental building blocks for superconducting electronics. As you will see in Chapter 7, it also explains some of the difficulties with processing high temperature superconductors.

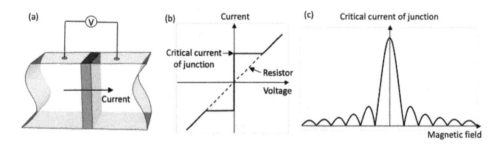

Fig. 6.12: (a) Diagram of a Josephson junction. (b) Current–voltage characteristics of a Josephson junction compared to a standard resistor. (c) Effect of magnetic field on critical current of a Josephson junction.

---

[6]These were named after a Welsh scientist called Brian Josephson who won the Nobel prize for physics in 1973 for his earlier discovery of the Josephson effect as a PhD student at the University of Cambridge.

[7]There is also an AC Josephson effect, but we do not need to know about that here. If you are interested, it relates to the fact that an alternating current is generated if you deliberately apply a fixed voltage across a Josephson junction.

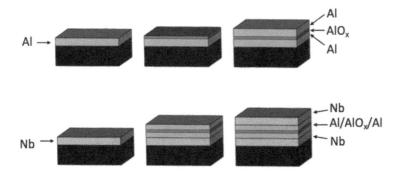

Fig. 6.13: Growth of Josephson Junctions.

So how do we go about making Josephson junction devices in practice? In fact, there are lots of different ways, depending on what material we are using and what kind of device we are making. What they all have in common is that the superconductor is almost always in the form of a thin film—a layer grown as a coating on top of a substrate of another material. It is very important that the growth of both the superconducting layers and the barrier layer are well controlled to make sure that the layers have the right thicknesses and chemistry. It is for this reason—the ease of growth of reliable Josephson junctions—that superconducting devices are often based on aluminium. In the conventional sense, aluminium is not a good superconductor. It has a critical temperature of 1.2 K compared to 9.2 K for elemental niobium, and its critical field is just 1% of that of niobium. Both metals can be grown relatively easily using standard thin film deposition processes like thermal evaporation or sputtering.[8] The major benefit of using aluminium is that it is pretty easy to grow a really nicely controlled aluminium oxide insulating layer on the top of it, just by introducing a bit of oxygen into the chamber during the Al deposition process. A sandwich of $Al/Al_2O_3/Al$ can therefore be grown just by opening and closing the oxygen gas valve at the right time (Fig. 6.13(a)). Although materials like Nb or NbN are much better superconductors, the growth of a nice niobium oxide layer turns out to be much more difficult than growth of aluminium oxide, so actually if Nb-based junctions are needed they are often made by depositing an Al-based junction stack in between the two Nb layers: $Nb/Al/Al_2O_3/Al/Nb$ (Fig. 6.13(b)).

6.5 SQUIDs

One of the neat applications of Josephson Junctions are in devices called SQUIDs. No, these have nothing to do with the rubbery sea creatures known for producing edible black ink. In fact they are Superconducting QUantum Interference Devices—devices that can pick up really tiny magnetic fields. In fact, these devices are in widespread use as magnetometers for measuring the magnetic properties of other materials! A DC

---

[8]Sputtering is a vapour deposition technique that uses energetic argon ions to physically knock atoms off the surface of a target material to vaporise them.

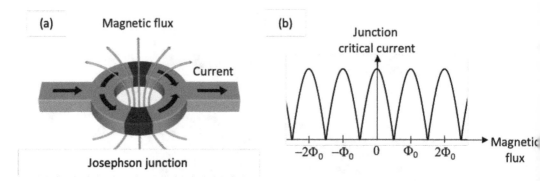

Fig. 6.14: (a) Diagram of a DC SQUID. (b) Critical current of the SQUID as a function of magnetic field.

SQUID consists of a ring-shaped superconducting track with two Josephson junctions at opposite edges (as shown in Fig. 6.14(a)). A DC current is passed from one side of the loop to the other. If the junctions are identical, the current splits, and half of it goes through one of the junctions and the other half goes through the other junction. The currents then meet up again at the other side of the loop. Provided the current going through each junction is below the critical current of the junctions, we would expect a constant superconducting current to flow through the device. The situation changes a bit when we put the device into a magnetic field. Now the two junctions are affected differently by the field, because the current we are passing through the device—let's call it the transport current—is going in a clockwise direction through one junction and an anti-clockwise direction through the other. When we apply a magnetic field, an additional circulating current will be generated in the loop that will be circulating in the same direction as the transport current through one of the junctions, but the opposite direction to the transport current through the other junction. The critical current of the full device will therefore be influenced by how much magnetic field passes through the loop. In fact, when the two separate supercurrents going through each of the branches of the device meet up again, they interfere with each other and the critical current of the device will oscillate as the magnetic field increases, as shown in Fig. 6.14(b). This is just the same as the interference fringes you get on the screen in a Young's double slit experiment using light waves. Since the spacing of the fringes in the SQUID turns out to be equal to a single magnetic flux quantum, it is possible to measure changes in magnetic field to the accuracy of a single superconducting flux quantum. This makes SQUIDs the most sensitive magnetic field sensors available.

SQUIDs can also be made from high temperature superconductors that can be operated at liquid nitrogen temperature. These have found applications for detecting valuable mineral deposits hidden deep underneath the ground using a technique called transient electromagnetic prospecting . Here, eddy currents are induced in an underground conducting mineral deposit by pulsing a current in a loop of wire on the surface, and sensing the magnetic field produced using the SQUID. Commercial systems have discovered deposits of precious metals like silver worth several billion dollars.

Under the Lens

SQUID sensitivity

SQUIDs are known to be the most sensitive sensors for magnetic fields. But how small a field can they detect? The field inside a SQUID can only exist in multiples of the magnetic flux quantum $\Phi_0 \approx 2 \times 10^{-15}$ Wb, as can be seen in Fig. 6.15 and so this is the minimum amount of flux that can be detected. Remembering that the magnetic B field is just the *magnetic flux density*, the minimum B field that can be detected depends on the area of the loop in the SQUID (A).

$$B = \frac{\Phi_0}{A} \tag{6.7}$$

A typical device may have a circular loop about 1 cm in diameter. This means that it has an area of $A = \pi(0.005)^2$ m^2.

$$B_{min} = \frac{\Phi_0}{A} \approx \frac{2 \times 10^{-15}}{\pi(0.005)^2} \approx 2.5 \times 10^{-11} \text{ tesla} \tag{6.8}$$

This magnetic field is six orders of magnitude lower than the earth's magnetic field. The best SQUID-based sensors have actually managed to detect magnetic fields as small as 10^{-18} tesla!

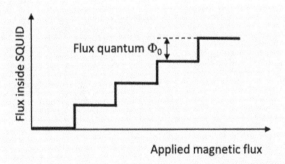

Fig. 6.15: Magnetic flux inside a SQUID loop as a function of external flux.

6.6   Superconducting computers

Josephson junctions have the attractive property of being able to suddenly switch from a zero voltage state to a finite voltage state just by increasing the current above the junction critical current. This is exactly what is needed in a computer which works by manipulating 'bits' of information, each of which can be either 'off' or 'on' (in the '0' or '1' state) at any particular moment. In a simple 'bit', based on a Josephson junction, the 'off' state corresponds to when the current is below the critical current of the junction so no voltage is produced, and the 'on' state corresponds to higher currents when there will be a voltage across the junction. Alternatively, the information can be encoded by whether or not there is a single quantum of flux contained within a loop of superconductor with a Josephson junction. The neat thing about making a computer out of a superconductor is that the switching speeds are really fast, so it is possible to make very high speed computers—about a hundred times faster than the current state-of-the-art desktop computers.

Another major advantage is that superconducting microprocessors consume much less energy than conventional semiconductor-based technology. You would be forgiven for thinking that computers do not use much energy compared to many other technologies like heavy industry and transport, but you may be surprised to learn that it actually accounts for about 1% of global energy consumption. It is responsible for the same amount of carbon dioxide emissions as the commercial airline industry. Not only that, but it is increasing at an alarmingly fast rate as the computational power of the devices increases. More than half of the energy they use is in transporting the electrical signals around from one place to another, between storage devices and logic gates for example, because the interconnections are conventional (resistive) metals. A superconducting computer would be able to send these signals around without any resistance, greatly reducing power consumption. Of course, there are many challenges to using superconductors, including the need to cool them down to low temperatures, so it is not likely that we will all be using superconducting laptops in the near future. However, it is not a crazy idea for the large data centres which are responsible for the majority of the energy consumption. That being said, superconducting electronics is a long way behind semiconductor technology in terms of the number of devices that can be integrated on a chip, with a maximum Josephson junction count being around 1 million in a 100 mm^2 chip compared with a transistor count of around 16 thousand times as many (16×10^9) in the latest Apple M1 chip which has about the same area! Moreover, Josephson junction devices are really promising candidates for a whole new type of computer—quantum computers—that use quantum effects to enable many calculations to be carried out in parallel, as will be discussed in Chapter 10.

Chapter summary

- The ability of a superconductor to carry resistance-free current is associated with the quantum nature of the electrons in the material. In fact, a superconductor is rather like a giant atom. In atoms the electrons go round in orbits without dissipating energy, just like the macroscopic persistent currents that circulate in a superconductor.

- In superconductors, electrons pair up to form Cooper pairs and the rules governing the motion of these pairs are different to the rules for unpaired electrons, leading to the dramatic change in macroscopic properties.

- In conventional low temperature superconductors, the attractive force between electrons arises because the electrons interact with the lattice of ion cores in the material. In high temperature superconductors the origin of the attractive force that leads to the electrons pairing up is still not clear, more than 30 years after their discovery, and this remains one of the most important questions that theoretical physicists are trying to answer.

- Quantum tunnelling—a purely quantum effect with no classical analogue—is exploited in real devices that can be used as very sensitive magnetic field sensors (SQUIDs) or for applications like computing.

- As we will see in the next chapter, the Josephson effect that is relied upon in these devices also presents a real headache for materials scientists wanting to make high temperature superconductors that can carry a lot of current.

7

Grain Boundaries: the Good, the Bad and the Ugly

If I was asked what the essence of materials science is, I would say that it is all about understanding how defects—bits of the material that are not quite perfect—affect the properties of a material. Making sweeping generalisations, engineers tend to work at a larger scale, where they are using the known properties of materials to figure out how to make a structure or device. Chemists usually work at the smaller scale of atoms and molecules to understand how substances interact with each other and react. Physicists work at all length scales, but they are looking to understand the fundamental principles of nature and therefore tend to work with 'perfect' or 'model' systems where they can simplify things. Materials science has its own special place, nestling in between these disciplines, and the detailed understanding of defects is at the heart of the subject. In particular, we are generally concerned with solid objects (known formally as *condensed matter*) and as we saw in Section 4.1, most solids are crystalline—they have a regular, repeating arrangement of atoms or ions. Imperfections in the regular arrangement are known as crystal defects and more often than not they play a vital role in determining the actual properties of a material. For example, we think of metals as being malleable. They can be bent and deformed relatively easily without breaking. That is how many metal objects are made, as we saw in Section 5.5 when we talked about extruding and drawing NbTi wires. However, if we had a perfect metallic crystal without any defects, we would not expect the material to be as malleable. Instead it would probably behave more like a ceramic, eventually breaking in a brittle fashion. This is because the way in which a metal permanently deforms is by the movement of millions of crystal defects called dislocations (see Section 5.3).

Let's take another example: silicon. Silicon is the element just below carbon in the periodic table and, in its elemental form, each silicon atom bonds covalently to four other silicon atoms making a three-dimensional giant structure just like diamond. As such, we would not expect it to conduct electricity very well. We actually categorise it as a *semiconductor* because only a relatively small amount of energy, heat for example, is needed to allow some of the electrons to move. One of the ways that we can make it more conductive is to replace a small fraction of the silicon atoms with phosphorus atoms. Since phosphorus is in group five of the periodic table, it uses four of its outer electrons to bond with the four neighbouring silicon atoms and has one electron left over in its outer shell that it does not need for the bonding. This electron is able to break away from the phosphorus atom (delocalise) and can travel through the material, carrying electrical current. For this reason, adding some phosphorus impurity to silicon helps to increase its electrical conductivity. It is the fact that we can precisely control the conductivity of silicon that it has become the basis of all our current computer technology as well as for things like solar cells that generate electricity from sunlight.

Grain boundaries are just one of the multitude of defects (introduced in Section 4.1) that superconducting materials will be riddled with, and most of them will have some effect on the superconducting properties. What makes grain boundaries special enough to warrant their very own chapter. in this book is that they can be both very good, and very bad, for the performance of the superconductor. In some materials we want loads of them, but in others we must avoid them at all costs! The purpose of this chapter is to explain why.

7.1   What are grain boundaries?

As I have already mentioned, the vast majority of all materials (including supercon-ductors) are crystalline. This means that the atoms arrange themselves in a regular pattern that repeats throughout the material. We call the repeating structure a *lattice* and the repeating unit is known as the *unit cell*. Let's take a material like sodium chloride (salt) as an example that you will be familiar with. In salt, the sodium and chloride ions are arranged in a cubic lattice—the unit cell is a cube. Each grain of salt is a single crystal which means that the unit cells are arranged next to each other in a perfect three-dimensional grid throughout the whole material—like a Rubik's cube before you start twisting it. The sides of the cubic cells point along the same directions everywhere in the crystal. If you look at salt grains carefully you will see that they are cubic in shape. This is just a manifestation of the fact that the atoms inside are arranged in a cubic structure. In this case, the atomic planes that make up the cube faces have lower surface energy than planes that cut through the crystal at different angles, so these are the ones that make up the surfaces of the macroscopic (large scale) crystal.

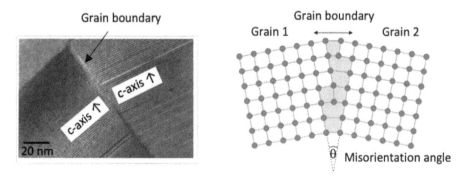

Fig. 7.1: (a) High resolution transmission electron micrograph showing the structure of a tilt grain boundary in Tl-1223 (Eastell *et al.*, 1998). (b) Diagram of a tilt boundary.

Fig. 7.2: Electron backscatter diffraction map showing the grain structure in a pellet of MgB_2(Courtesy of Z. Gao, University of Oxford.)

Unlike single grains of salt, most solid materials are not single crystals. They contain many different crystals, or grains, all joined together, and so we refer to them as being *polycrystalline* in just the same way that *poly*gons are many-sided shapes and *poly*mers are molecules made up of strings of many smaller chemical units. We refer to each crystal as a *grain*. Within each grain the atomic lattice is aligned along the same direction, but in neighbouring grains, the lattice is aligned in a different direction. Where the two grains join together we get a *grain boundary*. An example is shown in the high resolution electron micrograph in Fig. 7.1(a). We can understand how polycrystalline structures may form by considering the solidification of liquid metal. Suppose we take a container of molten metal and cool it down to just below its melting point. It does not solidify all in one go immediately. Instead, tiny clusters of atoms come together to form small solid volumes (we call them *nuclei*) within the liquid, which then grow outwards until they run into each other. Each nucleus forms a different grain and where they meet each other we get a grain boundary. Since each nucleus forms independently of its neighbours, their orientations are random. This is why we generally see misalignments in the crystal lattice at the grain boundaries. Because grains are three-dimensional objects, grain boundaries are two-dimensional—they are like the walls between rooms. We can control the size of the grains in a material using lots of clever materials science tricks—anywhere from a few nanometres (a few millionths of a millimetre) upwards. In fact, if you go out and look at some steel objects like lampposts you will probably be able to see by eye the grain structure of the zinc that is used as a protective coating to stop the steel from rusting. A nice image of the grain structure of a magnesium diboride superconductor is shown in Fig. 7.2.[1]

---

[1] The image is generated using a technique called *electron backscatter diffraction*, which is performed in a scanning electron microscope. Each colour represents a different crystal orientation.

Under the Lens

Low angle grain boundaries

Low angle grain boundaries actually consist of a stack of dislocations. For a tilt boundary like the one shown in this schematic (Fig. 7.1(a)), the dislocations are indicated by the 'T' symbols. They are edge dislocations that basically consist of an extra half plane of atoms squashed in at the top, and the dislocation itself is a line-shaped defect running into and out of the page where the 'T' is marked. In between the dislocations, the horizontal crystal planes are continuous across the two grains—with just a slight distortion, thus minimising the number of bonds that have to be broken to form the boundary.

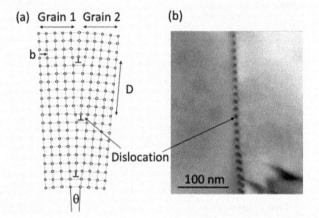

Fig. 7.3: (a) Low angle grain boundary consisting of a stack of edge dislocations. (b) A TEM image of a low angle grain boundary in MgB_2 superconductor. Reproduced from (Song and Larbalestier, 2004) with permission from Springer Nature. Copyright 2004.

We can use trigonometry to work out the grain boundary misorientation from the distance between the dislocations (D) if we know the *Burgers vector (b)* of the dislocations. The Burgers vector is just the horizontal length of the extra bit that has been added above the dislocation, and in the schematic, it is equal to the lattice parameter of the crystal (the side length of the lattice).

$$sin\theta = \frac{b}{D} \approx \theta \qquad (7.1)$$

Since θ is a small angle (it is, after all, a low angle grain boundary), we have used the small angle approximation $sin\theta \approx \theta$ (for θ in radians). If you haven't come across radians before, a complete circle has 2π radians, so radians can be converted to degrees by multiplying by the factor $\frac{360}{2\pi}$.

Let's estimate the misorientation angle of the low angle grain boundary in MgB_2 superconductor in the nice transmission electron micrograph in Fig. 7.1(b). Dislocations in MgB_2 usually have a Burgers vector equal to the a-axis lattice parameter of the hexagonal structure, so we will assume that the dislocations have $b = 0.3$ nm. Counting the number of dislocations on the image, we can see that we have about seven dislocations in a 100 nm length, so $D = \frac{100}{7}$

$$\theta = \frac{b}{D} \approx \frac{0.3 \times 7}{100} = 0.021 \text{ radians} \tag{7.2}$$

$$\theta = 0.021 \times \frac{360}{2\pi} \approx 1.2° \tag{7.3}$$

Therefore we estimate that the grain boundary in the image has a misorientation of only about 1.2°.

There are two different kinds of grain boundary: tilt boundaries and twist boundaries. Both of them are rotations of the crystal on one side of the boundary relative to the other side, but the difference is the axis of rotation. This can be easily demonstrated using hardback books (Fig. 7.4). Firstly, if you take a single book and open at a page somewhere in the middle, you have essentially made a tilt boundary. At a real grain boundary you do not end up with a gap like you do with the book—the crystals on the two sides extend until they meet up with each other. To see how a twist boundary works imagine putting two books in a neat pile, one on top of each other. Now twist the top one so it does not line up with the bottom one and you have got the equivalent of a twist boundary. In the vicinity of either of these grain boundaries, there is a narrow zone of material in which the bonding has been disrupted and the atomic arrangement is significantly different from rest of the crystal. Strain from this disruption will spread to an even wider zone than shown in Fig. 7.1(b). In general, the larger the misorientation angle of the boundary—the more tilted it is—the wider the disturbed zone at the boundary. Picture a grain boundary as a wall with two large dimensions and one narrow dimension. The misorientation angle (and type of material) influences how thick the wall is.

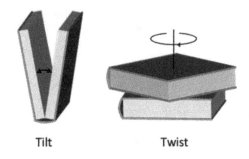

Tilt   Twist

Fig. 7.4: Illustration of the rotation axes of tilt and twist grain boundaries.

7.2 The good: flux pinning

Grain boundaries are regions where superconductivity will be weaker, and may even be lost altogether, because the bonding is disrupted locally. This means that they can act as good flux pinning sites. Remember that in Section 4.1 we introduced the idea that to get high currents the flux lines need to be pinned by defects to stop them moving and dissipating energy. Ideally we want the defects to be about the same size as the diameter of the flux line and this is determined by a fundamental property of the superconductor in question called its *coherence length (ξ)* which is roughly the distance between the electrons in a Cooper pair. The coherence lengths of some common superconductors are given in Table 7.1.

Table 7.1 Coherence lengths and upper critical field values of selected superconductors.

Material	Coherence length (nm)	B_{c2} at 0 K (T)
Nb	38	9
NbTi	6	14
Nb_3Sn	4	28
$YBa_2Cu_3O_7$	~ 2 (∥ab)	>100
$Bi_2Sr_2CaCu_2O_8$	~ 1 (∥ab)	>100
$Bi_2Sr_2Ca_2Cu_3O_{10}$	~ 1 (∥ab)	>100

As you can see, since grain boundaries are about 1 nm thick, the flux lines in the low temperature superconductors (Nb, NbTi, Nb_3Sn, MgB_2) will be significantly larger. This means that grain boundaries will not be very effective pinning centres for flux lines that are perpendicular to them. However, since the grain boundaries are much bigger in the other two dimensions, flux lines lying in the plane of the grain boundary will be very effectively pinned by it. Therefore, since the volume of intersection of the flux line with a grain boundary changes with its angle relative to the boundary, we would expect some grain boundaries to be better pins than others. Whilst we expect all of the flux lines to be pretty much aligned along the direction of the applied magnetic field, the grain boundaries in a polycrystalline material will be at random orientations. The total pinning force will be an average value calculated by taking into account the geometrical probabilities of the different orientation angles. The more grain boundaries there are—the more densely packed they are—the higher the pinning force per unit volume will be. This means that we really want small grains in our superconducting materials to maximise flux pinning.

The Wider View

Grain boundary strengthening

Grain boundaries are effective barriers to dislocation motion in metals. Since deformation occurs by dislocation motion, holding up the dislocations has the effect of strengthening the material. Although dislocations are rather similar to magnetic flux lines in superconductors in terms of their geometry, the details of how they interact with grain boundaries is rather different. Unlike flux lines, dislocations like to travel on particular *slip planes* in the crystal. If we deform a metal plastically, dislocations are generated in the grains and start to move along the slip planes, but when they reach a grain boundary it takes additional energy for them to change direction and continue into the second grain because they are not in the right orientation to slip in the next grain. The result is that the dislocations get stuck, temporarily at least, at the boundary. Subsequent dislocations travelling on the same slip plane then 'pile up' behind the blockage—just like a traffic jam, as shown in Fig. 7.5. Eventually, if we increase the stress enough, the dislocations will be forced across to the second grain, or a new source of dislocations will be activated in the second grain and slip will continue. The upshot is that grain boundaries raise the *yield stress* of the material. The smaller the grains, the more closely spaced the grain boundaries and the greater the strengthening effect. This can be described using an empirical equation—one that is arrived at by experiment rather than theory—called the *Hall–Petch relationship*.

$$\sigma_y = \sigma_0 + \frac{k_y}{\sqrt{d}} \tag{7.4}$$

Here σ_y is the yield stress, σ_0 is a materials constant that tells us the initial stress needed to get the dislocations moving before they start piling up, k_y is the strengthening constant and d is the average grain size.

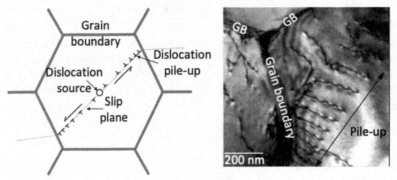

Fig. 7.5: Diagram and TEM image (courtesy of Dr Jack Haley, Department of Materials, Oxford) showing dislocation pile-ups at a grain boundary.

7.3   The bad: weak links

What about the high temperature superconductors? As you can see from Table 7.1, the coherence lengths of HTS materials ($YBa_2Cu_3O_7$, $Bi_2Sr_2CaCu_2O_8$, $Bi_2Sr_2Ca_2Cu_3O_{10}$ for example) are much smaller than those of the low temperature superconducting materials. In fact, they are rather similar to the thickness of a grain boundary. Grain boundaries will still be good flux pinning centres, for all the reasons already mentioned, but we have a major problem. In Section 6.4 we saw that layers of non-superconducting material sandwiched between slabs of superconductor act as weak links (Josephson junctions). If the non-superconducting layer is thin enough, superconducting current (Cooper pairs) can quantum-mechanically tunnel through the boundary without much problem, but as the layer gets thicker, the tunnelling current will decrease exponentially. So a thick barrier effectively cuts off the flow of supercurrent. But how thick is too thick? That depends on which superconductor we are talking about. The distance over which the supercurrent can tunnel without decaying too much is about the same as the superconductor's coherence length—the spacing of the electrons in the Cooper pair. This means that grain boundaries are pretty much invisible in low temperature superconductors because they are so much narrower than the coherence length that the current can tunnel straight through with no difficulty.

The same is not true for high temperature superconducting materials. Now, because their coherence length is about the same as the thickness of a grain boundary, we would expect to lose a good fraction of the current as it makes its way though the barrier. The grain boundary acts as a weak link. This is what we call the grain boundary problem in high temperature superconductors and it was first reported in a seminal paper by Dimos, Chaudhari and Mannhart (Dimos *et al.*, 1988). They made special thin film samples, each with a single isolated grain boundary to show that the critical current

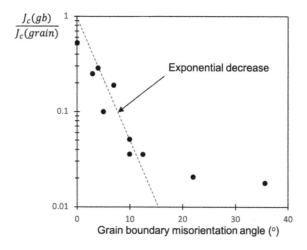

Fig. 7.6: Bicrystal study showing influence of grain boundary misorientation angle on J_c. (after (Dimos *et al.*, 1988).)

across a grain boundary exponentially decreases with misorientation angle, as shown in Fig. 7.6. Note that the y-scale of the graph is logarithmic. This means that although it looks like a straight line, the critical current actually drops off exponentially (much more rapidly) with increasing angle. In general terms, this is because the effective thickness of the grain boundary essentially increases with misorientation angle.

The upshot of the grain boundary weak link problem is that we cannot get large current densities through polycrystalline high temperature superconducting materials. We can really only afford to have very low angle grain boundaries in the sample, otherwise they will completely block current flow and the material will be useless. The problem gets even worse when we introduce magnetic field into the equation because the critical currents of Josephson junctions decrease drastically as we start to apply magnetic field (Fig. 6.12). This all means that if we are going to be able to use them in real applications we have to figure out ways of making high temperature superconductors in the form of pseudo single crystals. They do not have to be proper single crystals that have grown from a single nucleus, but we cannot afford to have any high angle grain boundaries. Any grain boundaries present have to have misorientation angles of less than about 10°. This is a huge processing challenge, particularly since we need to make many kilometres of wire if we want to make magnets. Impossible as it may seem—as you will see in Chapter 9—by the wonders of modern materials science, it can actually be done! A consequence of the grain boundary problem is that in high temperature materials we cannot rely on grain boundaries to do our flux pinning. We need to create flux pinning centres by other means.

What about other structural defects though? Does the weak link problem also mean that other structural defects of similar size are also bad? Actually no, and the reason comes down to their shape. High angle grain boundaries form barriers across the sample that the current cannot get across or around—they act like a dam stopping the flow of water in a river. Dislocations are more like posts or pillars; they are one-dimensional. Unlike a dam, if you stuck a post in the middle of the river, the water would just flow around it. Impurity particles are even better. They are a bit like rocks or pebbles in the river. You would have to have an awful lot of them to block the river flow completely!

7.4   The ugly: chemical segregation

The structural disorder—lack of perfect alignment of the atoms—at a grain boundary makes them particularly suitable places for impurity atoms to choose to reside. First of all think about the case where a small atom in the lattice is replaced by a larger impurity atom. The atoms around this site will all have to move outwards a little bit to give it room, and this costs energy in the form of elastic distortion (strain). We refer to the region where the atoms are squashed closer to each other than they would ideally like to be as a compressive strain field, and it typically extends further than just the nearest neighbour atoms. Although it may not be as obvious, if we replace a large atom in the lattice with a smaller impurity atom, we also generate a strain field. This time the surrounding atoms all have to stretch out a bit to help fill the gap, so we call this a tensile strain field and it costs us energy as well. At a grain boundary, the arrangement

Wire axis

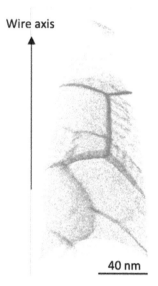

40 nm

Fig. 7.7: Atom probe tomography reconstruction showing Cu segregation to grain boundaries in Nb_3Sn. Cu atoms are represented by single orange dots. (Courtesy of Laura Wheatley, University of Oxford.)

of the atoms is already disrupted, which typically leads to some regions where there is a compressive strain field (atoms are too close together) and some regions where there is a tensile strain field (atoms are too far apart). The system can save itself energy overall if big impurity atoms sit in a region at the grain boundary where there is a tensile strain field, and small impurity atoms sit in regions of the grain boundary where there is compressive strain. This leads to *chemical segregation* in which the impurities tend to collect together at the grain boundaries (or other defects). The other reason that impurities often sit at grain boundaries is because they can provide a faster pathway for movement of the atoms within the solid—a process called *solid state diffusion*. This will be discussed in more detail in Chapter 8, but here it is sufficient to know that grain boundaries are a bit like motorways: they provide a network of fast routes through the sample that the impurity atoms can whizz along. This means there is both a thermodynamic (energy) driver and a kinetic reason for chemical segregation at grain boundaries.

Figure 7.7 shows grain boundary segregation of Cu and Ti atoms in Nb_3Sn supercon-ductor. The data has been obtained using a very specialist technique called *Atom probe tomography*. It works by stripping atoms layer by layer from a tiny specimen using high voltage or laser pulses. From the time-of-flight before hitting the detector, the mass-to-charge ratio of the ionised atoms stripped from the sample can be measured using mass spectroscopy. In addition, from the location where the ions hit a position-sensitive detector we can figure out their original position in the sample. This means that we can reconstruct the original sample, atom by atom, to get a three-dimensional picture of the distribution of the chemical elements. Unfortunately, this technique only

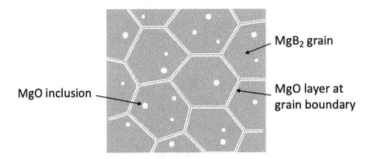

Fig. 7.8: Insulating MgO layer at the grain boundaries in MgB$_2$ and isolated MgO inclusions.

works if the specimen is in the shape of a very sharp needle, and most materials we want to measure are not naturally in that form. This means we first have to machine the samples into the right shape, often using a focused ion beam microscope so that we can choose exactly which bit of material to look at. If we want to look at chemical segregation at grain boundaries, atom probe tomography is the way to go.

So why have I so unfairly branded this section as 'The ugly'? I, for one, think that images like the atom probe tomographs shown in Fig. 7.7 are really beautiful. The ugliness refers to the effect that chemical segregation at grain boundaries can have on the properties of the material. In low temperature superconducting materials, a certain amount of chemical segregation may not be a problem. In fact, it can improve properties by helping to slow down grain growth. This is known as the *solute drag effect* because the grain boundary is pinned in place by the segregated impurities and so is less mobile, and it is one of our tricks for making materials with small grains. In some cases though, it can be a problem. Take the example of MgB$_2$. It is almost impossible to make completely oxygen-free MgB$_2$ because Mg is so reactive. If the resulting magnesium oxide just happily formed as isolated inclusion particles in the material, then they would probably act as nice flux pinning centres. Unfortunately though, there is a tendency for continuous layers of the oxide, as well as amorphous (non-crystalline) phases, to form preferentially at the grain boundaries. This makes the non-superconducting zone at the grain boundary thicker (as shown in Fig. 8.6) and, even though MgB$_2$ is a low temperature superconductor, this can cause a weak link type problem like that in the high temperature superconductors.

The Wider View

Grain boundary corrosion

One of the problems that metallic objects suffer from is *corrosion*—the gradual destruction of the material by chemical (or electrochemical) attack. The example that you will probably be most familiar with is the rusting of iron. Alloys like stainless steels have been developed by materials scientists to provide improved corrosion resistance. This is often achieved by adding a decent slug of chromium to the alloy (typically over 10%). Chromium is ideal because it reacts with oxygen in the air to form a thin, self-healing oxide layer that protects the metal underneath and stops it from reacting with the environment.

One of the undesirable things that can happen is that chromium carbides can form at the grain boundaries. Chromium is sucked from the immediate surroundings, leaving chromium-depleted zones in the vicinity of the grain boundaries. This makes the grain boundaries more susceptible to corrosion. Elements like niobium, titanium or tantalum may be added in these alloys because they form more stable carbides than chromium, reducing the chances of these chromium-depleted zones being formed.

Chapter summary

- Grain boundaries are highly desirable in low temperature superconducting materials because they are very good at trapping magnetic flux lines and so can increase the critical current density of the superconductor. This means that we actively want to make the grain size in low temperature superconductors as small as possible to maximise the density of grain boundaries.

- Grain boundaries in high temperature superconductors are really bad! This is because the small coherence length of the high temperature superconductors means that grain boundaries are thick enough to act as weak links and block the flow of the supercurrent.

- The higher the misorientation angle of the grain boundary, the worse the problem is. This means that we preferably want to make high temperature superconducting materials in the form of single crystals, or at least eliminate all of the high angle grain boundaries.

- In the wider context, controlling grain size is important in nearly all materials because grain boundaries can affect mechanical and chemical properties, as well as electrical properties like superconducting current transport.

8

Battles with Brittleness

As we saw in Chapter 5, the reason that niobium-titanium (NbTi) is by far the most widely used superconducting material is that it is ductile and so can be relatively easily drawn into wires. We have at our disposal a whole host of other materials that have superior superconducting properties, but they are all much more difficult and expensive to make. This means that they are only ever used in practice if the properties of NbTi are not good enough, either because higher magnetic fields or higher operating temperatures are needed. In fact, the desire to use NbTi wherever possible meant that instead of using a higher performance superconducting material the designers of the Large Hadron Collider opted to use NbTi for the magnets that bend the proton beam cooled with superfluid helium to lower the operating temperature to 1.9 K. However, if we need to get magnetic fields above about 10 tesla, as is the case for magnets for plasma confinement in nuclear fusion reactors, we cannot use this trick and we have to resort to something with a higher upper critical field (B_{c2}) like niobium-tin (Nb_3Sn) or even a high temperature superconductor. Alternatively, if we need our device to operate at a higher temperature we need to select a superconductor with a higher critical temperature (T_c). One possibility is the relatively cheap magnesium diboride (MgB_2) if we do not need high magnetic fields—otherwise we are looking at needing to use the super expensive high temperature superconductors. The problem with all of these higher performance superconductors is that they are *brittle* and that makes it very difficult to make wires and wind them into magnets. First of all we will tackle how to process the brittle low temperature superconductors MgB_2 and Nb_3Sn in this chapter, before proceeding to the even more challenging high temperature superconducting ceramics in Chapter 9.

8.1   Intermetallic compounds

In Section 5.1 we introduced the idea that niobium can be alloyed with titanium by randomly replacing some of the Nb atoms with Ti atoms to form a *substitutional solid solution*. The Ti basically dissolves in the Nb, keeping the body-centred cubic crystal structure essentially unchanged. We saw from the phase diagram 5.1 that at temperatures above about 600°C, we can put in any concentration of Ti we like and it stays in the same phase—it has complete solid solubility. In fact, this is rather unusual behaviour. As mentioned in Section 5.1, there are a set of empirical guidelines called the *Hume-Rothery rules* that tell us how likely it is that element X will dissolve in element Y to form a substitutional solid solution. The first is to do with the relative

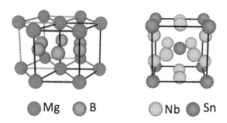

Fig. 8.1: Crystals structures of MgB_2 and Nb_3Sn.

sizes of the two different atoms—the size factor: the difference in the atomic radii of X and Y needs to be less than about 15% to form a solid solution over an extended range of compositions. The other rules tell us that it helps for X and Y to have similar crystal structures, valency and electronegativity.[1] Alloys that perform well on these rules tend to be ones, like NbTi, where the two elements are close together in the periodic table.

So what happens if the conditions outlined in the Hume-Rothery rules for substitutional solid solutions are not met? In these cases it is most likely that compounds form with fixed ratios of the two elements. Instead of the atoms of X randomly substituting for atoms of Y, ordered structures form where the X atoms sit on one site in the crystal lattice and Y atoms sit on a different site. The bonding between the atoms has some ionic character, but is still essentially metallic in nature and so they will conduct electricity in the solid state (albeit less well than pure elements like Cu or Al). However, the mechanical properties of these *intermetallic compounds* are very different to the solid solution alloys. In the elastic region (at low strains), intermetallics tend to be stiffer than pure metals because the degree of ionicity makes the bonds stronger, but as we will see, the most dramatic difference happens when the material is deformed to higher strains beyond the elastic limit. MgB_2 and Nb_3Sn superconductors are both intermetallic compounds, with MgB_2 having a hexagonal crystal structure consisting of layers of Mg and B atoms and Nb_3Sn having a cubic structure catchily called the 'A15' structure (see Fig. 8.1).

8.2 Brittle fracture

In Section 5.3 we discussed what happens when a ductile metal yields. Dislocations (linear crystal defects) start to move and produce permanent (plastic) deformation. But what happens in materials where dislocations are not so prevalent, or it is much more difficult to get them to move? In this case, pre-existing tiny cracks open up when the material is pulled in tension, and a stress concentration builds up at the tip, as shown in Fig. 8.2.[2] When the stress is large enough, the bonds at the tip of the crack break and the crack grows. Once it starts growing, the stress builds up rapidly (as

---

[1]Valency is essentially the number of outer electrons in the atoms, and electronegativity is the tendency of the atom to attract electrons when it forms a chemical bond.

[2]Remember a stress is defined as the force that is applied divided by the cross-sectional area carrying that load.

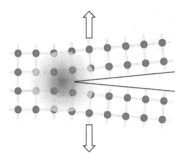

Fig. 8.2: Brittle fracture at a crack showing stress concentration at the tip.

there is less cross-sectional area of material to take the applied force) and so the crack very quickly propagates across the entire sample, breaking it completely in two. The material fractures. The stress that can be applied to the material before it fractures depends not only on the bond strength, but also on the maximum flaw size (the longest pre-existing crack) that is present in the material. Therefore, minimising flaw size can greatly improve a material's *fracture toughness*—its resistance to fracture. In fact, it is even possible to make objects like springs out of concrete if you process it properly to eliminate cracks. It is also the reason that tiny things like carbon nanotubes are heralded as being super strong. They are so small that the maximum flaw size is necessarily even smaller, making them more resistant to fracture.

The ordered crystal structures of intermetallic compounds like Nb_3Sn and MgB_2 with specific atoms taking particular sites in the lattice, make it more difficult to generate and move dislocations when a stress is applied. This means that they show limited ductility, and brittle fracture tends to occur instead. The two failure mechanisms are quite different, and lead to distinctive differences in the appearance of the broken surfaces of the material (Fig. 8.3). In ductile materials, fracture is preceded by necking—narrowing of the cross-sectional area of the sample. The fracture surfaces tend to show a 'cup and cone' shape where one of the broken pieces resembles a cup, and the

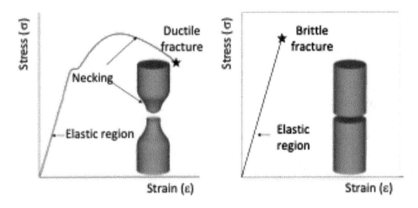

Fig. 8.3: Stress–strain curves and fracture surfaces of ductile brittle fracture.

other resembles a cone that fits inside the cup. In contrast, perfectly brittle materials in which no plastic deformation can take place, follow elastic (reversible) behaviour right up until the point where pre-existing cracks start to grow, leading to catastrophic failure. Fracture tends to occur straight across the material, perpendicular to the applied stress, making smooth, flat fracture surfaces. However, grain boundaries can be weaker points in some materials, leading to fracture following the shape of the boundaries rather than being completely flat.

8.3   Magnesium diboride superconductor

Magnesium diboride (MgB_2), although being rather simple chemically and rather conventional in terms of its superconducting properties, is the most recently discovered technological superconductor. By technological, I mean that you can buy it in the form of long lengths of wire that can be used to wind magnets or make power transmission cables. It has a higher critical temperature (39 K) than the niobium-based compounds (< 20 K), but it is nowhere near as high as the so-called high temperature superconductors (~ 90 K). Its use is limited to relatively low magnetic field applications though, because it does not have a very high upper critical field. The main attraction of MgB_2 is the low cost and good availability of the raw materials, as well as the relatively simple processing methods compared to the high temperature superconductors, coupled with its ability to operate at temperatures that are easily accessible with cryocoolers—special refrigerators that can reach about 20 K without needing liquid cryogens. So how do we go about making it in the form of the wires that are needed for most applications?

Powder-in-tube wire-making process

Obviously the brittle nature of intermetallic compounds like MgB_2 makes it impossible for them to be processed into long wires using a standard drawing process like the one used to make multifilamentary NbTi wires described in Section 5.5. The superconductor would break up into lots of separate particles and the superconducting path would no longer be continuous. Not very good for getting high currents through the wire! One way of overcoming these difficulties is to use a processing method known as *powder-in-tube*. Basically, you take a metallic tube—sheath—and pack it as tightly as you can with powder. The whole assembly is drawn down to the desired diameter wire and then it has to be *heat treated* to join up the powder particles. With MgB_2, there are two different flavours of this process that can be adopted known as *in situ* and *ex situ*. Let's start with in situ wires. They are called this because the MgB_2 phase is formed inside the wire after it has been drawn. The metal tube is packed with a mixture of unreacted magnesium and boron powders, the assembly is drawn to form a wire and then a chemical reaction to form the MgB_2 compound is carried out on the wire at about 700–900°C.

$$Mg_{(s)} + 2B_{(s)} \rightarrow MgB_{2(s)} \tag{8.1}$$

This process has several things in its favour. The first is that the reaction happens at relatively low temperature, helped by the fact that the magnesium melts. This is both cost-effective and minimises the chance of the metal sheath around the wire reacting and contaminating the superconductor. Secondly, because the MgB_2 grains are formed during the heat treatment, the grain boundaries are nicely fused together, making it easy for supercurrent to get through the material. But it has one major flaw. There is a substantial volume change when the chemical reaction takes place. The MgB_2 product phase takes up a smaller volume than the Mg and B reactant phases added together, so the microstructure is riddled with holes that we call *porosity*. This vastly reduces the cross-sectional area for superconducting current to flow though, limiting the critical current (I_c) of the wire.

Under the Lens

Porosity in MgB_2

The fraction of porosity produced in the reaction of Mg with B to form MgB_2 can be figured out from the relative densities of the reactants and products and the relative atomic masses of Mg and B. The first step is to calculate the molar volume (V_m) of each substance, given by $\frac{\text{mass per mole}}{\text{density}}$.

Substance	Mg	B	MgB_2
Density (g cm^{-3})	1.74	2.37	2.57
Mass per mole (g mol^{-1})	24.0	10.8	45.6
Molar volume (cm^3 mol^{-1})	13.81	4.56	17.74

Then, the total volume of the reactants and products can be estimated (per mole of Mg), taking into account their stoichiometric ratio seen from the chemical equation 8.1: $Mg:B:MgB_2 = 1:2:1$. This allows the volume reduction to be calculated.

$$\text{Volume reactants} = V_m(Mg) + 2V_m(B) = 13.81 + 4.56 = 22.93 \text{ cm}^3$$
$$\text{Volume products} = V_m(MgB_2) = 17.74 \text{ cm}^3$$
$$\text{Fraction porosity} = \frac{\text{Volume reactants} - \text{Volume products}}{\text{Volume reactants}}$$
$$= \frac{22.93 - 17.74}{22.93} = 0.23$$

This simple calculation shows that, from the volume change that occurs when Mg and B react to form MgB_2 we would expect to introduce as much as 23% porosity.

(1)   (2)   (3)

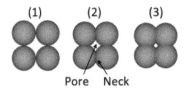

Pore   Neck

Fig. 8.4: Stages of the sintering process.

The alternative ex situ method involves packing the metal tube with pre-reacted MgB_2 powder. After drawing to form the wire, a heat treatment is performed to persuade the separate powder particles to fuse together. This is called *sintering* and because there is no liquid magnesium this time, it relies on *solid state diffusion*—the process by which atoms move around in a solid to find the most energetically favourable locations. Solid state diffusion is a much slower process than liquid state diffusion and so we typically need higher temperatures above 900°C for ex situ processing of MgB_2. The sintering process progresses in a series of stages illustrated in Fig. 8.4. Firstly 'necks' form where powder particles touch each other. These are just like our own necks—bits of material joining two particles (our head and our body) together, and gaps called pores are left in between the necks. This neck formation process is driven by the energy saving associated with reducing the surface area of the particles. As sintering continues, the pores gradually shrink and become more spherical in shape. It is often difficult to get rid of the pores completely unless you apply pressure during sintering, but the density of ex situ processed MgB_2 is still much higher than is possible with the alternative in situ process because we do not have the volume change associated with the MgB_2 formation reaction. This ought to make the critical current of the wires significantly better.

However, we have a different set of problems now, which can be explained by looking at the magnesium–boron phase diagram (Fig. 8.5). The MgB_2 phase region (highlighted in red) looks like a line on the diagram rather than an area because very little variation in chemical composition can be tolerated in this phase—it wants to have exactly two boron atoms per magnesium atom. For this reason it is called a *line compound*. Now look at what happens if you go up in temperature at the same composition (move vertically upwards on the diagram). You will see that there is no liquid zone. Instead the MgB_2 decomposes to MgB_4 and Mg vapour. Since this particular version of the diagram is for atmospheric pressure (1 atmosphere $\approx$ 100 kilopascals), this means that if you are at the temperature at the top of the MgB_2 region (at about 1550°C), you would expect MgB_2 to be in equilibrium with MgB_4 and 1 atmosphere of Mg vapour. At lower temperatures, we would still expect the decomposition reaction to occur, but the equilibrium pressure of Mg vapour (its vapour pressure) will be lower. You can understand this by thinking about what happens when you boil a kettle; steam starts to come out of the spout well before the water boils. It starts boiling properly (at 100°C) when 1 atmosphere pressure of steam is in equilibrium with the water. The chemical equation of this reversible process is

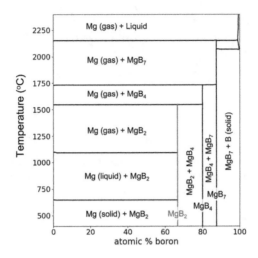

Fig. 8.5: Mg-B phase diagram.

$$MgB_{2(s)} \rightleftharpoons MgB_{4(s)} + Mg_{(v)} \qquad (8.2)$$

The tendency of MgB_2 to decompose throws up a problem in ex situ processing. In order for sintering to occur we typically need to use high heat treatment temperatures over 900°C because solid state diffusion gets much faster at higher temperatures, and at these temperatures thermal decomposition is considerable. So, if we want to use conventional ambient pressure sintering, we have to balance these two competing effects—the heat treatment will be a compromise between densification of the material and decomposition of the superconducting MgB_2 compound. We can push things in our favour though if we do the sintering process at higher pressure. You can understand why this works by looking back at equation (8.2). Le Chatelier's principle tells us that if we increase the pressure, the position of the equilibrium will shift in the direction that reduces the number of moles of vapour phase. In this case it means the equilibrium will move to the left—stabilising the MgB_2 as desired. Also, it turns out that densification is also easier at higher pressure. When we make bulk pellets of MgB_2 it is fairly easy to apply some pressure during sintering, but it is rather more difficult, though not impossible, with continuous lengths of wire. Even when we manage to make really dense MgB_2 it does not necessarily mean that it will be able to carry a lot of supercurrent. It turns out that, although the grains are packed closely together and grain boundaries are not inherently weakly linked (unlike in the high temperature superconductors), the boundaries that form between the original powder particles are often still not fused very well together and are riddled with a layer of insulating impurity phases that limit the ability for current to get through (Fig. 8.6). The upshot is that both in situ and ex situ processed MgB_2 have issues with electrical connectivity—how easy it is to get current through—but for different reasons. In situ material has a lot of porosity that limits the cross-sectional area for current trans-

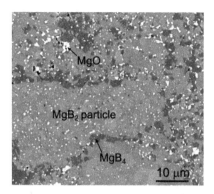

Fig. 8.6: Scanning electron microscope (SEM) image showing particles of MgB_2 surrounded by MgB_4 impurity grains (dark). Fine-scale MgO particles (white) can also be seen to decorate the outside of individual MgB_2 grains.

port, whereas in ex situ material the particle boundaries present a barrier to current transport.

MgB_2 can also be made in the form of bulk pellets for permanent magnet or shielding applications discussed in Chapter 4 using either an in situ or an ex situ technique by packing the powders into a mould instead of a metallic tube.

Optimising performance

In the ex situ technique, the higher the pressure that you apply during the sintering, the lower the processing temperature that is needed to make it densify. For bulk MgB_2 processing it is possible to use extremely high pressures of around 5 gigapascals (5×10^9 pascals), which enables the sintering process to be carried out in a few minutes at temperatures several hundred degrees lower than is needed for low pressure sintering. But how big is 5 gigapascals (GPa) in real money? Imagine turning the Eiffel Tower upsidedown on its tip—the pressure exerted on the ground would be about 5 GPa. Amazingly, there are industrial presses for making artificial diamonds in large quantities at these kinds of enormous pressures, so it is perfectly possible to access this kind of pressure in a commercial process. The benefit you get is that the lower processing temperature means that smaller grain sizes can be achieved because the solid state diffusion process that is needed for grain growth will be slower. Since grain boundaries provide most of the flux pinning in MgB_2, these small grain sizes are highly desirable for optimising superconducting performance. However, there are other problems with using these extreme pressures, such as increased processing cost and the material having a tendency to crack and fall apart when the pressure is released! Using starting powders with smaller particle sizes can also help because the smaller the initial grain sizes, the smaller the final grain sizes after processing.

Under the Lens

Mechanism of solid state diffusion

Solid state diffusion has been mentioned in passing several times already, so it is probably a good time to look at what is going on in a bit more detail. The idea of atoms in gases and liquids diffusing is probably already familiar to you. For example, you may have learnt that small molecules like oxygen and carbon dioxide can diffuse in and out of the cells in our body. You will also know from experience that if you put a drop of coloured dye into a glass of water it is not long before the whole lot has changed colour—diffusion at work again. What is happening in both of those situations is that there is a tendency for the molecules to move from a region of high concentration to a region of low concentration (down a concentration gradient). This process is driven by entropy: the dispersal increases the disorder of the system which lowers its free energy. The diffusion rate is given by an equation called *Fick's first law*, which relates the flux (J) of atoms or molecules to the concentration gradient ($\frac{dc}{dx}$) and a temperature dependent *diffusion coefficient (D)*. Here, the term 'flux' just means the number of atoms passing through an area of 1 m^2 in 1 second.

$$J = -D\frac{dc}{dx} \qquad (8.3)$$

Since gases and liquids are fluids, the atoms or molecules can move around very easily because there is lots of free space, but in solids it is not so easy. In fact, if you had a perfect crystal with no defects in it, solid state diffusion would be painfully slow because it would require two atoms simultaneously switching places. But we are lucky. Real crystals are defective, and for self-diffusion (e.g. Cu atoms diffusing in Cu) or the diffusion of substitutional impurities, one particular type of defect is our friend—the vacancy. This point defect was introduced very briefly in Section 4.1 and is just a missing atom in the crystal lattice. It provides a valuable hole that another atom can move into.

The process goes like this. Atoms in a crystal jiggle around all the time at a frequency of about 10^{13} vibrations per second, and sometimes one of these jiggles manages to kick an atom off its site and into a neighbouring vacancy. If the atom has hopped to the left, the vacancy has essentially hopped to the right, as shown in Fig. 8.7. Another atom can now swap places with the vacancy and so on. If there is no concentration gradient in the material, for each jump a vacancy is equally likely to travel to any of the neighbouring sites (i.e. left or right, up or down, in or out, in a cubic lattice). This means that the net flux of vacancies is zero. However, if there is a concentration gradient in the material, the rates of forward and backward motion are no longer equal and there will be a net flux of vacancies and atoms in the opposite directions. The diffusion rate depends on the diffusion coefficient of the material, which in turn is influenced by the number of vacancies there are and how easily they can move.

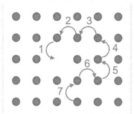

Fig. 8.7: Vacancy mechanism of solid state diffusion.

So how does the diffusion coefficient depend on temperature? Well, actually there are two different factors to consider. The first is that, to get from one atomic site to its neighbour, the atom has to get over an energy barrier (it has to squish through a gap). This is just like the activation barrier in the collision theory of chemical reactions that you may have come across before. The probability of an atom managing to get over the activation barrier for migration (E_m) goes up with increasing temperature (T) in a way that scales with the Boltzmann factor $e^{-\frac{E_m}{k_B T}}$ where k_B is the Boltzmann constant. Secondly the number of vacancies in the sample also increases with temperature in a similar way. This is because it costs a certain amount of energy to create a vacancy—its formation energy E_f—and so the more thermal energy available, the higher the chance that a vacancy will be created. The number of thermally generated vacancies is therefore proportional to $e^{-\frac{E_f}{k_B T}}$. Putting all of this together, we find that the diffusion coefficient varies with temperature according to the following equation.

$$D = D_0 e^{-\frac{E_m}{k_B T}} e^{-\frac{E_f}{k_B T}} = D_0 e^{-\left(\frac{E_m + E_f}{k_B T}\right)} = D_0 e^{-\frac{E_a}{k_B T}} \tag{8.4}$$

Here D_0 is a temperature-independent materials constant and $E_a = E_m + E_f$ is the activation energy for diffusion.

The general equation that gives the temperature dependence of reaction rates is called the *Arrhenius equation*, and is written as $k = Ae^{\frac{E}{k_B T}}$ where k is the rate constant of the reaction. It is arguably one of the most important equations in materials science. Interestingly, since the Boltzmann factor gives the statistical probability that a process can overcome an energy barrier, it is also important in understanding things like the temperature dependence of the magnetisation of a paramagnetic material and the conductivity of semiconductors.

Another way to improve the current-carrying properties of MgB_2 is by adding impurities—doping—usually with carbon-based compounds, but lots of different things have also been shown to be effective. Carbon tends to dissolve into the MgB_2 crystal lattice by substituting for boron atoms. This actually decreases its critical temperature, but it raises the upper critical field substantially making it more useful for magnet applications. The same thing happens to B_{c2} when Ti is added to Nb, and we call it making the superconductor 'dirty'. Doping with silicon carbide is particularly interesting because it has a dual purpose. Firstly the silicon carbide decomposes releasing carbon atoms to substitute in the MgB_2 lattice and increasing B_{c2} and secondly the remaining silicon atoms react with any Mg vapour being released from the decomposition of MgB_2 and forms fine particles of Mg_2Si which act as additional flux pinning centres. Bonus!

Applications of magnesium diboride

Owing to the relatively low upper critical field of MgB_2 it can only be used for power applications such as transmission cables, fault current limiters and machines like motors and generators, or low field magnets for MRI. Nevertheless, the relatively low cost of this material does make it attractive for niche applications. The problem is that, after the heat treatment, you have to be pretty careful with the wire so that the filaments do not break. There are two ways magnets could be made: 'wind and react' or 'react and wind'. The names tell you what they mean! In the first case, the magnet or device is wound before the heat treatment is performed, whereas in the second case the wire is fully processed and is then wound into the magnet. The second approach is the more desirable for magnet designers because it avoids having to heat treat a full magnet structure and gets round the problem of how to include the necessary insulation materials that would not withstand the high temperature. However, after the heat treatment, the wire has to be handled much more carefully, and cannot be bent into such a tight radius any more without breakages occurring.

There have been lots of demonstrator magnets and machines made using MgB_2, but here I will limit the discussion to one particular neat application for the high-luminosity upgrade of the Large Hadron Collider that is due to be up and running by the end of 2027. The idea of this upgrade is to increase the number of collisions per second that is possible in the collider, thereby increasing the number of Higgs bosons (for example) that can be measured per year from about 3 million to about 15 million and will allow physicists to understand the fundamental nature of these particles much better. To do it, some of the NbTi magnets will be replaced with stronger Nb_3Sn magnets to generate better focusing and to shorten the length of the bending magnets to make space for collimator devices that will narrow the beam. But where does the MgB_2 come in? Well, as the strength of the magnetic field increases, there becomes a vital need to move the power supplies for the magnets further away to stop them being destroyed. The current will now need to get from the power supply to the magnets over an extended distance of around 60 metres. And we are talking seriously big currents—well over 100,000 amps! That is not an easy task, and MgB_2 superconductor is ideally suited for it. The team at CERN has already built a couple of these MgB_2 cables in flexible helium gas cryostats that have successfully carried 54,000 amps. In fact, these cables hold the record of being the most powerful electrical transmission lines that have ever been built and operated!

8.4 Niobium-tin

So we have seen that MgB_2 superconductor is useful for some low field applications where a cheap conductor that can operate at higher temperatures than NbTi is desirable. Although Nb_3Sn superconductor serves a completely different purpose—making stronger magnets—a lot of the processing challenges are shared with MgB_2. It too is a brittle intermetallic that means it cannot be made into wires very easily. Unlike NbTi conductor, Nb_3Sn is still being actively developed and researched, driven by new high field applications like fusion reactors and particle accelerators. It has been around for over 60 years, and yet we are still finding out new things about the material and exploring new ways of making it and optimising its performance.

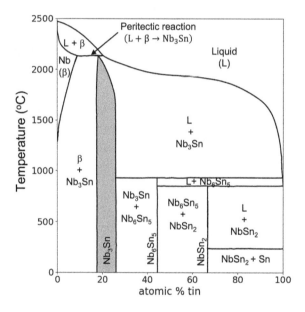

Fig. 8.8: Nb-Sn phase diagram.

A good place to start when we are trying to understand any new material is to look at the chemical phase diagram (Fig. 8.8). You will probably immediately notice that it is way more complicated than the Nb-Ti diagram we looked at in Section 5.1. However, we can ignore most of it. The superconducting Nb_3Sn phase is shown in grey. If you look directly above it on the phase diagram, you will see a region where liquid is in equilibrium with a different solid phase that is mainly niobium. This means that if you want to make Nb_3Sn from the melt, those solid niobium particles will have to react with the liquid to make the correct Nb_3Sn phase. This kind of reaction, where liquid is reacting with one solid phase to form a different solid phase, is called *peritectic solidification*, and they cause all sorts of difficulties for materials processing from the melt. One such problem is that it is difficult to get the reaction to go to completion because the new solid tends to form in a layer around the original solid, getting in the way of the reaction with the liquid. The other thing that you might notice is that the temperature of that peritectic reaction is well over 2000°C, so any kind of melt-process is not really possible for a commercial wire.

Wire processing

Until recently, there were three different approaches used commercially to make Nb_3Sn wires. One of those was the in situ powder-in-tube technique that was discussed in the context of MgB_2. In this case, $NbSn_2$ and Cu powders are packed into a niobium tube and reacted at high temperature to form the superconducting Nb_3Sn phase, but commercial production has recently been phased out because it is more expensive and does not quite manage to match the critical current densities achieved by the other routes. The second method essentially involves drilling a series of holes in a copper billet and inserting Nb rods and Sn rods. The entire assembly is drawn down to its final wire dimensions with the ductile Nb and Sn still separated from each other by the copper cladding. The wire is then heat treated at a temperature of around 650°C which allows interdiffusion (mixing) and reaction of the Nb and Sn to form Nb_3Sn. Because the tin rod is usually placed in the centre, the method is generally referred to as the *internal tin* process. The final commercial route for making Nb_3Sn wires is called the *bronze* process. In this case the Sn source is provided by using bronze (a Cu-Sn alloy) as the cladding material instead of Cu, and no extra Sn rods are needed.

Here we are only going to talk in detail about one of the many variants of the internal tin method that is produced commercially called the Resacked Rod Process (RRP®). The starting billet is shown in Fig. 8.9(a). As you can see, there is a large rod of Sn in the centre surrounded by a layer of Cu assembled using hexagonal cross-section Cu rods. Nb rods, each clad in a layer of Cu are arranged around this core before the whole assembly is put in an Nb diffusion barrier and an outer Cu tube. These assemblies are drawn down into wires and then 'restacked' in the arrangement shown in Fig. 8.9(b) before being drawn to the final wire dimensions.

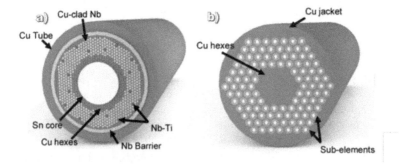

Fig. 8.9: The RRP® process: (a) a single subelement and (b) assembly of subelements. Reproduced from (Sanabria *et al.*, 2018) with permission from IOP Publishing. All rights reserved.

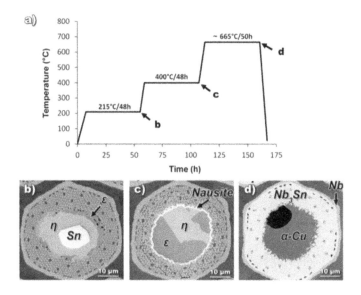

Fig. 8.10: Standard heat treatment of Nb$_3$Sn wire. Reproduced from (Sanabria *et al.*, 2018) with permission from IOP Publishing.

The heat treatment is absolutely crucial for getting wire that can carry a lot of current, and it turns out that the copper plays a crucial role. The process is typically done in several stages, as shown in Fig. 8.10(a), each of which has a particular purpose. First of all, a low temperature process at around 215°C is used to allow tin from the core to diffuse into the surrounding copper layer forming a Sn-Cu intermetallic compound called the 'eta' (η) phase.[3] This stage of the heat treatment is not strictly necessary as it turns out, but it is part of the traditional commercial process. Some residual tin is left in the centre of each subelement. In the second stage of the heat treatment, the temperature is raised to about 400°C and an Nb-Cu-Sn intermetallic compound called *nausite* forms in a ring about 0.5 micrometres in thickness forms where the layer of niobium filaments starts. Controlling the formation of this nausite layer turns out to be an important part of the process, partly because the copper cladding surrounding the individual niobium filaments tends to diffuse inwards across the nausite layer and into the core, leaving just niobium (and voids) around the outside. On heating further, the η phase in the core melts and large faceted nausite grains form at the interface. These decompose via a series of intermediate compounds and eventually, at 665°C a layer of Nb$_3$Sn forms.

$$nausite \to NbSn_2 \to Nb_6Sn_5 \to Nb_3Sn \tag{8.5}$$

During this process, tin diffuses outwards from the η phase core, finally leaving just Cu in the core.

---

[3]Remember the convention that we use Greek letters to name different phases to avoid confusion with different elements.

Optimising performance

As is the case with MgB_2, flux pinning in Nb_3Sn is mainly provided by grain boundaries. Much of the work on refining the wire-making heat treatments is with the aim of increasing the fraction of fine-grained Nb_3Sn that forms during the reaction. One of the ways people have found of doing that is to try to limit the amount of the Nb_6Sn_5 phase that forms because it seems to facilitate the growth of larger grained Nb_3Sn which carries lower currents. This can be done by increasing the ratio of copper to tin in the starting materials, and is one of the reasons why the powder-in-tube process which necessarily has a lower copper content never quite matched the performance of the other types of wire. Also, in the same way that carbon doping was introduced in MgB_2 to make it dirty, impurity dopants can be added into Nb_3Sn to increase its upper critical field. Typically the dopants of choice are titanium or tantalum which (mainly) substitute for niobium atoms in the Nb_3Sn lattice and can increase B_{c2} at 0 K by as much as 3 tesla! In the RRP® process described here, this is achieved in practice by replacing some of the niobium rods in the original subelements with NbTi or niobium-tantalum alloy rods. During the high temperature heat treatment, the dopants manage distribute themselves throughout the Nb_3Sn very effectively. In addition, dopants like hafnium and zirconium are being explored for another reason. Provided there is sufficient oxygen around (which there usually is), these reactive metals oxidise and form nanoprecipitates (Fig. 8.11). These actually have two benefits for flux pinning: the ones that are distributed within the Nb_3Sn grains act as great artificial flux pinning centres, and the ones that are sat at grain boundaries slow down grain growth during the heat treatment by reducing the mobility of the grain boundaries. The smaller the grains in the final material, the stronger the grain boundary flux pinning will be. This idea of adding impurity particles to slow down grain growth and achieve what is called 'grain refinement'—smaller grains—is a standard materials scientists trick.

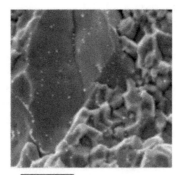

500 nm

Fig. 8.11: Ta-Hf-doped Nb_3Sn. Nanoparticles (white) can be seen at the grain boundaries and within the grains. Reproduced from (Tarantini *et al.*, 2021) under the Creative Commons Attribution 4.0 International license http://creativecommons.org/licenses/by/4.0/.

ITER fusion reactor

There are currently four main applications of Nb_3Sn superconductor: high field research magnets, high field NMR, state-of-the-art accelerator magnets (like the high luminosity LHC upgrade) and magnets for nuclear fusion machines, discussed in more detail in Chapter 10. Here, I will confine our discussion to talking specifically about the ITER fusion reactor that is currently under development in Europe. This is not intended to be a machine that will produce usable energy, instead being a huge demonstrator project to prove the viability of the technology. It is an incredibly long-term project that started its conceptual design phase in 1988 and the first plasma is not scheduled until December 2025.[4] After that, it is estimated to be another 10 years before a proper fusion reaction will take place. The superconducting magnets are needed to confine and control the plasma. In total, there are 48 different magnet coils in the ITER design, 18 of which are massive D-shaped Nb_3Sn toroidal field coils that are responsible for the confinement (Fig. 8.12a). Between them they store an eye-watering 41 gigajoules of energy—almost exactly the same amount of energy as a 5000 tonne weight dropped from the top of the world's tallest tower, the Burj Khalifa in Dubai (828 m high)! By comparison, the Large Hadron Collider stores a puny 10.5 gigajoules of energy in the magnetic field. Sickeningly, the atomic bomb that exploded above Hiroshima in the second world war released energy equivalent to the magnetic energy stored in about 365 ITER fusion reactors.

Although ITER is still a way off from being fired up, the specifications of the superconducting wire needed for the magnets was fixed decades ago and by no means stretches the ultimate capabilities of modern Nb_3Sn wires. Even so, there were unforeseen problems with the test coils and conductors which did not perform as well as they were expected to, and a special project had to be carried out to find out why. It turns out that the problem was with the degradation of performance of the conductor when it

Fig. 8.12: (a) First ITER toroidal field coil completed in Europe. (b) Transporting a toroidal field coil to the ITER site. (Courtesy of ITER.org.)

---

[4]A plasma is state of matter that consists of a gas of ions and free electrons that is electrically conductive.

was stretched (under strain). The Nb_3Sn for ITER was made by nine different suppliers. In fact, until ITER, the global production of Nb_3Sn was about 15 tonnes per year, and to make the 600 tonnes needed for ITER, production had to be stepped up by an order of magnitude. Instead of specifying exactly what manufacturing method should be used, all of the wire had to meet a strict set of quality criteria that included things about their superconducting performance, their physical geometry, their ratio of copper to superconductor, and so forth. That meant that each supplier designed special ITER strand using their favourite in-house method—the only way to have secured the amount of conductor that was required.

It is also worth mentioning that the Nb_3Sn ITER coils are made using the 'wind and react' method. This means that those enormous D-shaped coils are wound before the Nb_3Sn phase has been formed to avoid breakages, and the entire thing has to go through that precise thermal treatment! They then have to be transported very carefully to the construction site in the south of France by sea, barge and finally by road on enormous trucks (Fig. 8.12b).

Chapter summary

- Niobium-titanium is the only technological superconductor that can be fabricated by any standard metallurgical wire-making process. All of the other materials that have better superconducting properties are brittle, making them more difficult to process and use.

- The general strategy is to heat treat the wires after they have been drawn (and maybe even after they have been wound into magnets), either to react precursors into the superconducting phase or to encourage existing particles of superconductor to fuse together (sinter). These methods enable the formation continuous filaments of superconductor capable of carrying high currents.

- The difficulty with processing means that NbTi is nearly always used in practice, unless higher field or higher temperature operation is required. MgB_2 wires can be made relatively cheaply because the raw materials are not expensive, but Nb_3Sn wires are always considerably more expensive than NbTi.

9

High Temperature Heroics

The twentieth century saw the discovery of a whole host of metals and alloys that exhibited the remarkable phenomenon of superconductivity. By the 1980s, the record critical temperature had steadily increased from 4 K, just four degrees above absolute zero, to about 30 K. On top of that, in the 1950s a great advance was made in terms of understanding how superconductivity works on the microscopic level with the quantum 'BCS' theory proposed by Bardeen, Cooper and Schrieffer, for which they went on to receive a Nobel Prize in 1979. The theory essentially assumes there is some force of attraction between electrons which allows them to overcome the electrostatic repulsive force and form Cooper pairs that can travel through the material without experiencing resistance. As discussed in Section 6.2, it was found that this attraction originated from how the electrons interact with the vibration of atoms in the crystal lattice which we can think of as another kind of particle called *phonons*. This put a theoretical upper limit on the critical temperature at around 30–40 K, and so by the 1980s it was believed that superconductivity was fully understood and there was little further progress to be made.

That all changed dramatically in 1986 when superconductivity was discovered in copper-oxide-based compounds—known as *cuprates*—for the first time (Bednorz and Müller, 1986). Within a few months, scientists around the world were playing with tweaking the chemistry by doing things like replacing barium with strontium to increase the critical temperature. Spurred on by experiments that showed that applying pressure to the compound caused a substantial increase in critical temperature, the scientists tried to emulate this effect by replacing the lanthanum ions with smaller yttrium ions to strain the lattice and exert *chemical pressure*. The new compound—commonly called YBCO—was found to superconduct at 93 K, completely smashing the BCS limit of 30 K. This major breakthrough led to an explosion of activity and was heralded as the start of a new era in which superconducting technology would become commonplace.

However, over 30 years on, these *high temperature superconductors* (known as *HTS* for short) have still not lived up to the hyperbolic promises made in the initial aftermath of their discovery. The main reason is that, although the fundamental superconducting properties are amazing, the materials themselves are extraordinarily difficult to work with. It has simply taken a lot longer than initially expected to make the long lengths of wire capable of carrying the large currents that are essential for large scale applications.

As you will see in this chapter, through some remarkable materials science innovations, high temperature superconductors are now poised to start delivering on their initial promise.

9.1   The cuprate compounds

Following the initial discovery of superconductivity in La-Ba-Cu-O, it was found that this particular compound was not a one-off. A vast number of similar compounds also superconduct. The common feature in all of these materials is that if you look at the atomic scale inside the crystals, they all have layered structures consisting of planes of copper and oxygen atoms separated by blocks containing the other elements, shown schematically in Fig. 9.1. It is the copper-oxygen planes that are responsible for super-conductivity, so they can be thought of as conduction planes. The blocks separating them are essentially insulating. Their role is to act as a reservoir for charge, enabling the number of electrons in the copper-oxygen conducting planes to be tweaked. The cuprates are divided into families based on their chemistry and are given names that tell us what elements are present. For example, YBCO (pronounced 'ibco' or 'Y-B-C-O') relates to a crystal containing yttrium ($\underline{Y}$), barium ($\underline{B}a$), copper ($\underline{C}u$) and oxygen ($\underline{O}$), and BBSCCO (pronounced 'bisco') consists of bismuth ($\underline{B}i$), strontium ($\underline{S}r$), calcium ($\underline{C}a$), copper ($\underline{C}u$) and oxygen ($\underline{O}$).[1]

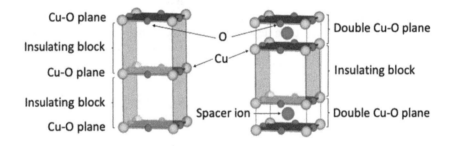

Fig. 9.1: Generic cuprate superconductor crystal structures with single and double copper-oxygen conduction planes.

Moreover, within each family there is a series of different compounds that can be made, each with a different number of copper-oxygen planes. Each family is a *homologous series* and can be defined by a general equation. You may have come across this idea before if you have learnt some organic chemistry. For example, the *alkanes* are a homologous series of hydrocarbons with the general formula C_nH_{2n+2}, where n can be any whole number: $n = 1$ is methane (CH_4), $n = 2$ is ethane (C_2H_6), $n = 3$ is propane (C_3H_8) and so forth. *Alkenes* are a different homologous series with a different general formula C_nH_{2n}. For the superconducting compounds, these formulae look more complicated because they contain more elements. Taking the BSCCO family as an example, the general formula for the homologous series is $Bi_2Sr_2Ca_{n-1}Cu_nO_{2n+4}$,

---

[1] For simplicity, the initials of the element names rather than their full chemical symbols are used.

where n is defined as the number of copper-oxygen planes in the crystal and can take values of 1, 2 or 3, although the $n = 1$ compound does not superconduct. To specify which of these compounds we are referring to without having to give the full chemical formula, we give each one a nickname based on the first element and the ratio of the cations (positive metal ions). For example, $Bi_2Sr_2CaCu_2O_8$ is called 'bismuth 2-2-1-2' (Bi-2212 for short) because it has 2 Bi ions, 2 Sr ions, 1 Ca ion and 2 Cu ions. By the same token, $TlBa_2Ca_2Cu_3O_9$ is called Tl-1223 for short. Tl is the chemical symbol for thallium—you may have heard of it as 'The Poisoners' Poison' because it has been responsible for several infamous murders. In most of the cuprate families, critical temperature increases with the number of copper-oxygen planes in the structure, as shown in Table 9.1. YBCO, or more generally (RE)BCO—where RE represents a rare-earth element such as Y, Gd, Sm, Eu etc.—is the slight exception, because only versions of the compound containing two copper-oxygen planes are found to superconduct. Although the (RE)BCO materials may seem simpler at first glance because they contain fewer elements (four instead of five), in terms of their crystal structures they are quite complicated because there are copper atoms in the insulating blocks as well as in the copper-oxygen planes.

Table 9.1 High temperature superconducting cuprates.

General formula	n	Formula	Shortened notation	T_c (K)
$Bi_2Sr_2Ca_{n-1}Cu_nO_{2n+4}$	1	$Bi_2Sr_2CuO_6$	Bi-2201[2]	<20
	2	$Bi_2Sr_2CaCu_2O_8$	Bi-2212	85
	3	$Bi_2Sr_2Ca_2Cu_3O_{10}$	Bi-2223	110
$Tl_2Ba_2Ca_{n-1}Cu_nO_{2n+4}$	1	$Tl_2Ba_2CuO_6$	Tl-2201	<80
	2	$Tl_2Ba_2CaCu_2O_8$	Tl-2212	108
	3	$Tl_2Ba_2Ca_2Cu_3O_{10}$	Tl-2223	125
$TlBa_2Ca_{n-1}Cu_nO_{2n+3}$	1	$TlBa_2CuO_5$	Tl-1201	<50
	2	$TlBa_2CaCu_2O_7$	Tl-1212	80
	3	$TlBa_2Ca_2Cu_3O_9$	Tl-1223	110
	4	$TlBa_2Ca_2Cu_4O_{11}$	Tl-1234	122
$HgBa_2Ca_{n-1}Cu_nO_{2n+2}$	1	$HgBa_2CuO_4$	Hg-1201	94
	2	$HgBa_2CaCu_2O_6$	Hg-1212	127
	3	$HgBa_2Ca_2Cu_3O_8$	Hg-1223	133
$(RE)Ba_2Cu_{n+1}O_7$	2	$YBa_2Cu_3O_7$	Y-123	92
	2	$GdBa_2Cu_3O_7$	Gd-123	94
	2	$DyBa_2Cu_3O_7$	Dy-123	95
	2	$EuBa_2Cu_3O_7$	Eu-123	95
	2	$NdBa_2Cu_3O_7$	Nd-123	96

9.2   Chemical fussiness

When we talked about niobium-titanium superconductor in Section 5.2 we saw that we can change chemical composition of the alloy over a wide range without making much difference to the critical temperature of the alloy; pure niobium has almost the same critical temperature as an alloy with 50% of the niobium atoms replaced with titanium atoms. This could not be further from the truth in the high temperature superconductors. Instead of being fairly tolerant to changes in chemistry, they are incredibly fussy. To get the highest transition temperatures we have to get precisely the correct proportion of each of the chemical species, and the atoms must all be sitting in exactly the right positions in the crystal. The reason basically comes down to the fact that it is absolutely vital that these materials have precisely the right number of electrons in the copper-oxygen planes. Moving too far away from this optimum will reduce T_c drastically.

If we take Bi-2212 as an example, we would expect from the charges on each of the ions in the compound that we would need 8 O^{2-} ions ($8 \times (-2) = -16$) to balance the 2 Bi^{3+}, 2 Sr^{2+}, 1 Ca^{2+} and 2 Cu^{2+} ions ($(2 \times 3) + (2 \times 2) + (1 \times 2) + (2 \times 2) = +16$) and achieve charge neutrality. If we have exactly this ratio, we say that the compound is *stoichiometric*. However, perfectly stoichiometric HTS compounds do not superconduct at all. Instead, they turn out to be electrical insulators and exhibit a different kind of magnetism called antiferromagnetism.[3] To turn them into superconductors, we usually have to reduce the total electron count in the copper-oxygen planes a little bit. We can do this in several ways such as replacing some of the O^{2-} ions with F^- ions, or by removing some of the O^{2-} ions altogether leaving spaces behind called vacancies that have an effective positive charge. We call this process *chemical doping*. You have probably heard the term doping used in relation to athletes taking performance enhancing drugs. The idea here is essentially the same—a chemical is introduced in small quantities to enhance performance. Chemical doping is also done in semiconductors like silicon to improve electrical conductivity, as mentioned previously in Section 7.

In the most common high temperature superconductors, it turns out we need the copper-oxygen planes to be deficient in electrons—hole-doped—to achieve the right conditions for superconductivity. If we have too few holes, the superconductor is said to be 'underdoped'. However, if we make too many holes by removing too many oxygen atoms for example, the sample will become 'overdoped' and we will lose superconductivity again. This is because the superconducting state has to compete with other thermodynamic states, and it only wins (has the lowest energy) over a small range of compositions. If we draw a phase diagram to show what states are stable as we change doping and temperature, like the one in Fig. 9.2, there is a dome-shaped region in which superconductivity is stable. At the top of the dome the transition temperature has its highest value, and the superconductor is optimally doped. In the region above

---

[3]Antiferromagnetism is another form of magnetic ordering in which neighbouring atomic moments align in the opposite directions.

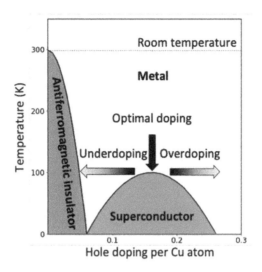

Fig. 9.2: Phase diagram of a generic cuprate superconductor.

the superconducting dome, the material is essentially metallic in nature, although it behaves in a rather strange way that we do not need to delve into here.

Exacerbating the issue of the extreme chemical sensitivity of the properties of HTS materials is the problem that they are not very stable. What I mean by this is that they react chemically with almost everything, including water vapour in the air, and this will degrade the superconducting properties. Most metals, except the noble metals like silver, platinum and gold, have a tendency to react with oxygen. Therefore, even in the best-case scenario if you coat a high temperature superconducting material with a metallic layer, oxygen atoms are likely to be sucked out of the surface of the superconductor and into the metal, leaving the superconductor with too little oxygen. This problem gets even worse because the HTS materials are ceramic oxides and need to be processed at high temperatures where chemical diffusion (the movement of atoms) is fast. However, we cannot avoid metals altogether if we want to make wires that can be wound into magnets or used as transmission power cables because ceramics are inherently brittle and they would simply break.

9.3 The grain boundary problem

In Section 7.3, we saw that grain boundaries—the planar defects between neighbouring crystals—act as barriers to the flow of a superconducting current in high temperature superconducting materials. They behave a bit like dams in a river. In low temperature superconductors this is not the case, because the electrons in a Cooper pair are much further apart than the thickness of a grain boundary and so they do not 'see' the defects and can flow straight across them. However, in HTS materials, the electrons bound in Cooper pairs are only separated by a distance of around 1 nanometre, which is similar to the thickness of non-superconducting (or weakly superconducting) material at a grain boundary. Low angle grain boundaries, where the crystals on either side of the

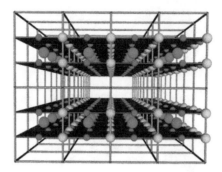

Fig. 9.3: Perfect alignment of the copper-oxygen planes looking along the wire direction.

boundary have only slightly different orientations, are effectively thinner and therefore have less of an impact than high angle grain boundaries. For this reason, in order to get high critical current densities (J_c values) in HTS materials, the general principle is to avoid high angle grain boundaries. On top of this, we know that HTS compounds have layered structures. Since it is the copper-oxygen planes that are responsible for superconductivity, we can get much higher currents flowing in a direction parallel to these planes than at right angles to them where the current would have to make its way across the insulating blocks in the crystal structure. This *anisotropy* —different properties in different directions in the crystal—means that we really want the crystals to be lined up with their copper-oxygen planes along the length of the wire (as shown in Fig. 9.3) for them to be able to carry large currents.

In general the practical approaches to overcoming these difficulties to making high current density wires can be split into two types known as first generation and second generation technologies. Not surprisingly, first generation wires were developed first. They are still used for processing BSCCO wires, but turned out not to be very successful for (RE)BCO materials (such as YBCO). For these materials, we have to resort to the more complicated and expensive second generation technology.

9.4   The BSCCO buddies

The Bi-Sr-Ca-Cu-O family is the proud host of not one, but two of our six technological superconductors—ones that can be bought commercially in the form of wires at the time of writing. They each have their own advantages and disadvantages. Bi-2223, with three copper-oxygen planes in each unit cell, has a considerably higher critical temperature than its shorter sibling, Bi-2212, which only has two copper-oxygen planes. In fact, Bi-2223 is a real contender for high temperature applications like power transmission cables because its T_c value of 110 K is more than 30 degrees above the boiling point of liquid nitrogen (77 K). Bi-2212, on the other hand, with a T_c value of around 85 K, is pretty useless in terms of its current carrying capacity at liquid nitrogen temperature because 77 K is so close to its critical surface.[4] There is very little headroom in terms of applying field or passing current before superconductivity will be lost. So you might wonder why Bi-2212 is still in the game. Well, in lots of applications, such as high field magnets, we do need to operate at liquid nitrogen temperature. What we want from our superconductor is the ability to still be able to carry lots of current in magnetic fields above 20 tesla—where Nb_3Sn stops working. We can live with having to cool to 20 K or even to 4 K if it allows us to access higher magnetic fields. These lower operating temperatures are far enough below T_c that the performance of Bi-2212 is not so different to Bi-2223.

Not only that, but Bi-2212 has a couple of other tricks up its sleeve that Bi-2223 can only dream of. The first of these is to do with its chemistry. Although both compounds contain five different elements, it turns out that it is much easier for the atoms to arrange themselves into the Bi-2212 compound than into the Bi-2223 compound. This means that wire processing is simpler and we are less likely to be left with a whole load of other impurity phases that we do not want. Bi-2212's other claim to fame is that it is the only high temperature superconductor that has been made successfully in the form of multifilamentary wires with circular cross-sections—round wires. Bi-2223 wires have to be flattened for reasons that will soon become apparent, and (RE)BCO materials (like YBCO) have to be made as a thin layer on top of a metal tape using second generation technology. This may not sound like a big deal, but round wires make life a lot easier for magnet designers because the properties are the same in all directions (isotropic). This is particularly important in applications like NMR and MRI where uniformity of the magnetic field is paramount. In addition, for many applications like accelerator magnets, it is desirable to weave many wires together into cables that can carry over 10 kiloamps of current, and this is much easier to do with round wires than with tapes.

Powder-in-tube method

The powder-in-tube method is a great way of turning brittle ceramic materials into fine wires. As we saw in Section 8.3, it is the standard method for making wires out of MgB_2 superconductor, as well as being one of the methods for making Nb_3Sn wires. As its name suggests, it involves packing powder of the superconductor or its constituent phases into a metal tube and then using standard metallurgical processes to gradually reduce the cross-section of the wire until the final dimensions are obtained. We can stop part-way through the process, roll the wire into a hexagonal cross-section, cut it into separate lengths and stack them together in a honeycomb arrangement inside another tube. The whole arrangement can then be drawn again to make a multifilamentary wire. We can repeat this process again if we need to end up with even more filaments in our final wire.

The problem is, even if we start by using powder of the superconducting compound, the powder particles will not be properly fused to one another, so it will not be possible

---

[4]The critical surface was defined in Section 1.3 as the surface in temperature, magnetic field and current density space gives the extent of operating conditions for which the material will be superconducting.

to pass a superconducting current along the length of the wire. There are several ways we can try to join the powder particles together after the wire has been made: sintering, solid-state reaction or melt-processing . The first was described in Section 8.3 and involves starting with the correct compound and heating it up to about $\frac{2}{3}$ of its melting point so that the atoms can rearrange themselves, effectively gluing the particles together. This sintering method is used to make ex situ MgB_2 wires. The second process, solid state reaction, starts with a mixture of precursor powders that are reacted together to form the desired product after the wire has been drawn using a heat treatment. This is the method used for in situ MgB_2 and powder-in-tube Nb_3Sn, and it tends to be an effective way to get good joints between the particles, but it can result in lots of holes (pores) if a gas is evolved in the process. The final possibility, melt-processing, involves starting with powder of the compound you want to make, but heating it up to a high enough temperature so that it melts, or partially melts. This is the best way to get good connectivity along the length of the wire, but it can be difficult to control the chemistry to make sure all the liquid solidifies back into the right compound. Melting cannot be used for MgB_2 or Nb_3Sn because the first decomposes to form a gas instead of melting, and the second would require temperatures over 2000°C. However, as you will see, melt-processing is a good method for making Bi-2212 wires. A slight complication is that because all high temperature superconductors react with, and are poisoned by, most metals, we have to use silver for the tubes instead of, say, stainless steel that is typically used in MgB_2 wires. This is slightly unfortunate because silver is not known for being cheap or strong. Using specialist silver alloys can help improve the strength, but there is not much we can do about the high cost!

Complex chemistry

Before we get on to talking specifically about how wires of Bi-2212 and Bi-2223 are actually made, we need to dip into the truly formidable chemistry of the Bi-Sr-Ca-Cu-Sr-O system. Luckily, we do not need to understand the intricacies for our purposes, so we will shamelessly gloss over the unnecessary details. As we have seen, the starting point for materials scientists wanting to understand the chemistry of a system is to go to the phase diagram—and here we meet our first challenge. For binary systems like NbTi which have two components (Nb and Ti), we can draw useful two-dimensional diagrams which have composition on the horizontal axis (with pure Nb at one edge and pure Ti at the other edge) and temperature on the vertical axis. It is nowhere near as easy to visualise the phase diagram when we have five different elements to consider! Not to be deterred, we can simplify the situation slightly by thinking of the components already being oxides rather than elements, reducing the problem to merely four dimensions at each temperature and pressure. We can draw this by putting each of the constituent oxides (Bi_2O_3, SrO, CaO and CuO) at a different corner of a tetrahedron, as shown schematically in Fig. 9.4. The main phases that we are interested in are the two superconducting compounds themselves (Bi-2212 and Bi-2223), the series of compounds known as alkaline earth cuprate (AEC)[5] which sit

---

[5] Alkaline earth elements are the metals in group II of the periodic table.

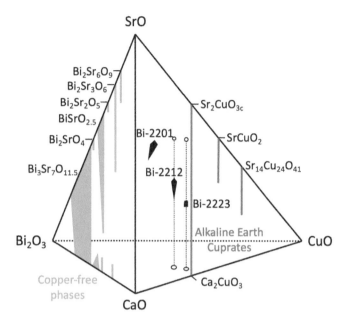

Fig. 9.4: Simplified BSCCO phase diagram, indicating the compositions of the most important phases. (after (Majewski, 2000).)

on the SrO-CaO-CuO face of the tetrahedron, the copper-free phases which sit on the Bi_2O_3-SrO-CaO face of the tetrahedron and the Bi-2201 phase. Although we may expect from the zero in its nickname that Bi-2201 contains no calcium, in fact it is possible for some of the strontium in this phase to be replaced with calcium because both elements are in group II of the periodic table. It therefore occupies a small volume of the tetrahedral phase diagram rather than just sitting on the Bi_2O_3-SrO-CuO face. We can also see a considerable difference between the Bi-2212 phase and the Bi-2223 phase in terms of how much space on the diagram they occupy. Bi-2212 is very tolerant, allowing strontium atoms to sit on calcium sites and vice versa, whereas Bi-2223 is much more fussy. To put some numbers on it, the Sr/Ca ratio can vary between about 0.5 and 6.5 in Bi-2212, but only in the narrow range of 1.9–2.2 for Bi-2223 (Duperray and Herrmann, 2003). The consequence is that it is much more difficult to home in on the right conditions for making pure Bi-2223. In order to stabilise the 2223 phase and make it easier to access, it helps a lot if we replace 10–20 % of the bismuth atoms with lead (Pb) atoms to make a phase that is formally referred to as (Bi,Pb)-2223. In fact, because it is so difficult to make Bi-2223 without the aid of lead, when we talk about Bi-2223 wires in this book we will always mean the leaded version.

Melt-processing of Bi-2212 wires

Unlike MgB_2 that decomposes and Nb_3Sn that has a very high melting point, the Bi-2212 compound can relatively easily be synthesised by a melt-processing route. Essentially, if you start with Bi-2212 solid powder and heat it up to about 890°C it decomposes into a mixture of solid phases, liquid and oxygen gas. The precise make-up of the solid phases can vary, but it typically contains a mixture of alkaline earth cuprates (like $SrCuO_2$) and Bi-2201. When we cool down the partial melt, the reverse reaction happens and the idea is that the liquid reacts with the solid phases to turn back into the superconducting Bi-2212 phase. However, in practice, the purity of the final product depends on various things including the strontium-to-calcium ratio, the oxygen gas pressure and the precise details of the heat-treatment. For example, if the Sr to Ca ratio is too high, very large Bi-2201 crystals form in the melt, and it takes an awfully long time to convert them back to Bi-2212 when we cool the mixture down. In contrast, by starting with the stoichiometric Sr:Ca ratio of 2:1, we can stop Bi-2201 forming in the first place, so the transformation back to Bi-2212 is much quicker and the final product contains less impurity phase. But we have to be careful; if we decrease the strontium content too much, another problem occurs and a pesky alkaline earth cuprate phase grows in the melt and hinders the conversion back to Bi-2212.

Figure 9.5 shows a generic heat treatment for the melt-processing of Bi-2212. It starts with ramp up to a temperature a little bit above the melting point of the Bi-2212, where it is held for a certain amount of time to make sure that the Bi-2212 has fully decomposed. The mixture is then cooled very slowly back through the melting

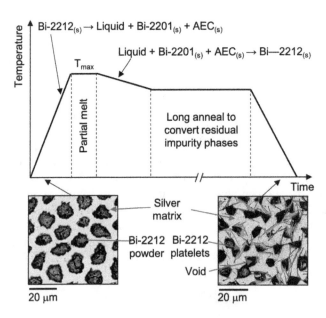

Fig. 9.5: Heat treatment for melt-processing of Bi-2212, with scanning electron micrographs showing cross-sections of the wire before and after the heat treatment.

point so that the Bi-2212 reforms. The slower the cool down, the larger the resulting Bi-2212 grains will become because there is lots of time for them to grow. This is ultimately good for getting a well-connected structure that can carry lots of current. Unfortunately, if we make the cooling rate really slow, the non-superconducting solid phases that are present in the melt will also grow, making it more difficult to get them to convert back to Bi-2212 and reducing the purity of the product. There are some tricks that can be played to try to balance these things out by carefully controlling cooling rate as a function of temperature.

That being said, it is inevitable that we will have some unwanted impurity phases as well as the superconducting Bi-2212 when solidification is complete. To try to transform as much of these residual unwanted phases as possible into Bi-2212, we do not cool it all the way down to room temperature straight away. Instead, we leave it soak at elevated temperature to enable the atoms to diffuse around in the solid state to find their equilibrium positions. This holding stage is called *annealing*, and even after about three days at around 830°C we will still not have managed to entirely get rid of the impurities!

So we can see that the melt-process, if done carefully, can lead to a pretty decent fraction of Bi-2212 grains that are well connected to each other. This is a great start, but on its own it will not be enough for long lengths of wire to be able to carry large currents. Ideally, we want all of the Bi-2212 grains to be nicely lined up with their copper-oxygen planes extending along the entire length of the wire. As it turns out, because of its layered crystal structure, Bi-2212 grains grow much faster in the plane of the layers than perpendicular to them, so they naturally grow in the shape of tiny little plates called platelets. These platelets form on the inside surface of the silver tube that surrounds each filament, with a natural tendency for the platelets to lie flat against the silver surface with their copper-oxygen planes lined up in roughly the right direction, as shown schematically in Fig. 9.6. The platelets often grow sideways so much that they extend into the silver matrix forming the spiky features that can be seen in cross-sectional images of the wire in Fig. 9.5. Although the grain alignment is not perfect, it is good enough that the current-carrying properties are still pretty good. In fact, it has recently been found that it is the presence of large holes inside the filaments that limits the current-carrying performance of Bi-2212 melt-processed wires, rather than the grain boundaries (Larbalestier *et al.*, 2014). So how can we get rid of the holes? A group from Florida State University has come up with a neat way to do it. Instead of doing the heat treatment in flowing oxygen at atmospheric pressure, they do it at higher pressure, which has the result of preventing the formation of the bubbles and increasing the density of the wire. The results are staggering: the engineering critical current density (which takes into account the whole cross-sectional area of the wire including the silver and the superconductor) is found to increase by as much as eight times when the high pressure treatment is used. At 4.2 K and high magnetic fields, this new Bi-2212 round wire outperforms all its competitors. If we could make it with fewer high angle grain boundaries and fewer impurity phases, it should perform even better!

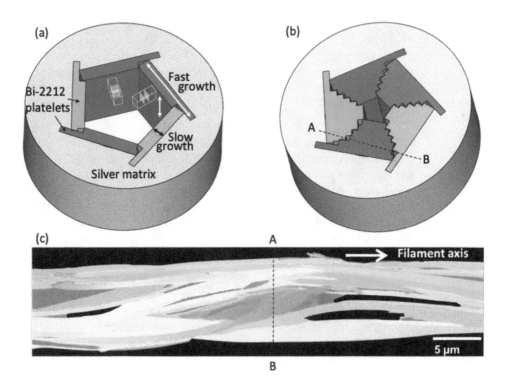

Fig. 9.6: Bi-2212 wire filament (a) an early stage, and (b) at a later stage in the solidification process. The crystal alignment of some of the Bi-2212 platelets is indicated in white, and arrows show the fast (white) and slow (black) growth directions. The different colours of the platelets indicate that the Bi-2212 grains have different crystallographic orientations. (c) Electron backscatter diffraction map of a longitudinal section of a filament showing the different grain orientations in different colours (courtesy of D. Larbalestier).

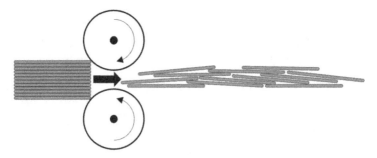

Fig. 9.7: Effect of rolling on a layered crystal like Bi-2212 with weak bonding between the layers.

Thermomechanical processing of Bi-2223 wires

Unlike Bi-2212, Bi-2223 is not so accommodating and cannot be successfully made using a handy melt-process. Instead, we do an in situ solid state reaction to make the Bi-2223, after the wires have been made. The precursor powder that is packed into the silver tubes is a partially reacted mixture of the constituent oxides, with the main component being our friendly Bi-2212 phase. Small amounts of other oxides are also present, bringing the overall cation ratios to the right values for Bi-2223. Essentially the process then involves doing a mainly solid-state reaction at about 830–840°C to form Bi-2223. However, unlike with the Bi-2212 wires, we do not have the luxury of the natural grain alignment that occurs during solidification. Instead, to get the Bi-2223 grains to line up in the right way, it turns out to be absolutely essential to flatten the wire by rolling using a giant version of a pasta machine before we do the reaction. But how does this work? Well, as we know, the Bi-2212 crystals in the precursor powder have a layered structure and the bonding between the layers is much weaker than the bonding within the layers—just like in graphite. This means that the layers have a tendency to slide over one another when we apply a shear force (for example by passing it through rollers), making plate-like crystals that lie down flat as shown in Fig. 9.7. In graphite, it is this sliding of the layers that makes it such a good lubricant. When we heat up the rolled wire to do the reaction, the Bi-2223 grains have a tendency to copy the alignment of the flattened Bi-2212 grains that they are growing from. The alignment is not perfect, and just like in Bi-2212, it is plagued by the formation of bubbles which further reduces the critical current. So, to get better grain alignment and to reduce the porosity, it is typical to use two or three extra rolling stages at intervals during the lengthy heat-treatment. The need for mechanical rolling (or pressing) to achieve grain alignment means that it is impossible to make a round cross-section Bi-2223 wire that is capable of carrying high currents.

9.5 Revolutionary (RE)BCO

Although we have seen that it is possible to make BSCCO wires and tapes which can carry high current densities, the need to use silver as the surrounding material to avoid chemical reactions introduces a couple of problems. Firstly, silver is an intrinsically expensive material, and whatever we do to make processing of the wires as cheap as possible, we cannot get around the fact that the raw cost of the silver is going to be crippling. The second problem with silver is that it is not known for having good mechanical strength, so BSCCO wires are not that well suited to high field magnet applications where the tensile forces generated are pretty huge. The BSCCO compounds also suffer from problems related to their extreme anisotropy—the properties being very different in different directions—which makes it very difficult to get really good flux pinning.[6] Even though (RE)BCO compounds like YBCO still have layered structures, their anisotropy is not so extreme as the bismuth-based compounds. This makes flux pinning a bit easier, but it has the unfortunate consequence that the layers in (RE)BCO no longer slide over each other as easily. This means that we cannot use mechanical tricks to persuade the grains to line up the way we want them to. Although

---

[6]The reason is complicated, but comes down to flux lines having more exotic structures.

it is, in principle, possible to melt-process these compounds, and that is the way they are made in bulk pellet form (see Section 9.7), the grain alignment that results from a powder-in-tube type approach is simply not good enough to get high current densities. The reason is probably a combination of it not being as easy to get good grain alignment mechanically and the compound being a bit fussier about high angle grain boundaries.

The consequence is that we have to use a different cunning ploy that involves producing a template—which we call a *substrate*—with the correct crystal alignment for the (RE)BCO crystals to copy when they are grown on top. One of the neatest ways to make a suitable substrate is to use the natural tendency for certain metals, like nickel, to develop nicely aligned grain structures when they are rolled into a thin tape and then heated up a bit. The rolling stage generates loads of dislocations inside the material as the metal is deformed plastically. When it is gently heated up, the defective nickel grains tend to regrow into defect-free ones to lower their free energy. This process is called *recrystallisation*. Nickel has a cubic crystal structure—the crystal is made up of an array of tiny cubic unit cells—and when we recrystallise it, the new grains like to form with their cubic cells all lined up in rows along the rolling direction (as shown in Fig. 9.8. When we have this kind of grain alignment we call it *cube textured* or *biaxially textured*[7] (where the word texture does not refer to surface roughness, but rather how the grains are aligned). Metal substrates produced in this way are rather sweetly referred to as *RABiTS*—Rolling Assisted Biaxially Textured Substrates.

Now the trick is to get the superconducting layer to grow on top of this nicely prepared template, with its crystals copying the same alignment. There are lots of different ways of putting down the superconducting layer, either from a gas or from solutions, but we cannot put it straight on top of the nickel substrate because nasty chemical reactions

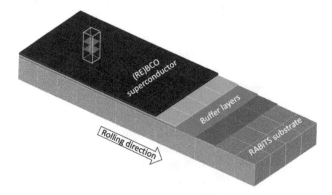

Fig. 9.8: RABiTS coated conductor with the orientation of a (RE)BCO unit cell shown in white. (Not drawn to scale).

---

[7]Biaxial here means that the cubes are aligned both within the plane of the tape and perpendicular to the plane.

will happen. In fact, it turns out that we have to put down several intermediate oxide *buffer* layers to transfer the texture of the substrate successfully to the (RE)BCO. The (RE)BCO is then capped with a silver coating to protect it, and the whole multilayered tape is coated in copper or brass to give it extra electromagnetic stability. We call the materials *coated conductors* because they have been made by coating one layer after the other. The final product is usually about 100 microns (0.1 mm) thick with a 1 micron (RE)BCO layer, and is available in lengths up to about 1 km. Even though the superconducting layer is only about 1% of the tape thickness, coated conductors can carry very high current densities because the (RE)BCO is essentially a 1 km long single crystal with no high angle grain boundaries! The problem with the RABiTS approach is that the nickel substrates that have to be used to get the required grain alignment are not great in terms of their mechanical strength and are usually ferromagnetic, which is a bit of a headache for some applications. To circumvent these problems, an alternative technique has been developed that uses an angled ion beam to force one of the oxide buffer layers to grow with the right alignment instead of relying on using the metal substrate as a template. This frees us up to use stronger, non-magnetic metal alloys for the flexible substrate, like a NiCrMo alloy or stainless steel.

The main problem that coated conductors have faced over the past couple of decades is that the multilayer deposition process that is required to get the necessary grain alignment is very expensive. The tape geometry is also not ideal for magnet engineers because it is more difficult to wind and its properties are very dependent on the direction of the magnetic field relative to the surface of the tape. Materials scientists have made lots of progress with developing cheaper processes using solution-based methods that do not need to be done under vacuum, as well as pushing up the current-carrying properties by improving flux pinning. A good example of this is the addition of a sprinkling of zirconium oxide to the starting materials for the (RE)BCO layer. During the high temperature processing, some of the barium reacts with the zirconium oxide producing barium zirconate ($BaZrO_3$). Under the right processing conditions, these can take the form of nanoscale rods that naturally align themselves perpendicular to the surface of the tape. They essentially make an arrangement inside the superconducting layer that looks a bit like the bristles of a brush. These one-dimensional defects are great for pinning flux lines and thus increasing the critical current density, provided the field is in a direction perpendicular to the tape surface so that the flux lines line up with the nanorods (see Fig. 9.9). Under different processing conditions, it is possible to make the barium zirconate form as more spherical nanoparticles instead, which provide more isotropic pinning. Many other artificial pinning centres—ones that are made by added impurity phases—also work, including rare-earth oxides.

Many other crystal defects that occur naturally during the (RE)BCO growth can also act as effective pinning centres. One common example is a two-dimensional defect called a *stacking fault*, where the stacking of the atomic layers goes wrong. Unlike grain boundaries, which must be avoided in high temperature superconductors, these stacking faults are always parallel to the direction that the current is flowing in, so they do not act as a barrier in the same way. Instead, they are really good at pinning flux lines that are parallel to the tape surface. To optimise the performance of coated

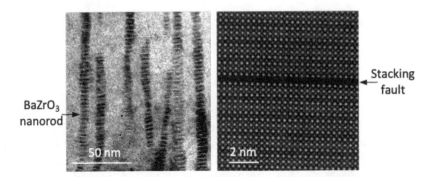

Fig. 9.9: TEM images showing barium zirconate nanorods and stacking faults in (RE)BCO coated conductor manufactured by SuperPower. (Courtesy of Y. Linden, University of Oxford.)

conductors for a particular application involves combining the natural and artificial defects so they work in synergy.

9.6   Flux creep

Up until now, we have talked about two different conditions for a type II superconductor in the mixed state: at currents below the critical current density the flux lines are pinned and do not move at all, and above the critical current density the flux lines are not pinned and they move freely, dissipating energy and generating resistance. As discussed in Section 4.1, we define the critical current density of a type II superconductor as the maximum current density before resistance is generated, rather than the much higher current density that would be required to actually pull apart the Cooper pairs and destabilise the superconducting state. In fact, the transition between these two states—pinned flux lines and unpinned flux lines—is not as clearcut as you might imagine. One way of thinking about it is that a strongly pinned flux line is sitting at a location where it is in a local energy minimum because there is a crystal defect of some sort. When we pass a high enough current through the material, there is a magnetic force—the Lorentz force—generated, which tries to push the flux line sideways away from this low energy site. If the Lorentz force is strong enough the flux line will succumb and move out of its energy minimum.

However, there is another way of getting flux lines to move away from their pinned sites. Imagine an array of low energy flux pinning sites, with energy barriers in between them (shown schematically in Fig. 9.10). If there is enough energy available, it may be possible for a flux line to get across the energy barrier and move from one pinning site (energy dip) to the neighbouring one. The most obvious source of this energy is heat. OK, so I know what you are thinking—superconductors have to operate at very low temperatures, so there cannot be very much heat energy available. You are right, but for high temperature superconductors the energy barriers may not be very high, and the higher the temperature we operate at, the more likely it is that the flux lines will move. Flux line motion by this mechanism is called *thermally activated flux flow* or *flux creep* by analogy with other thermally activated processes that occur in materials,

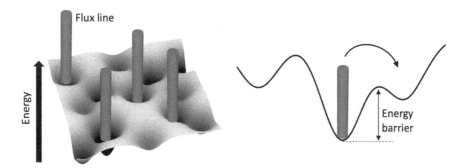

Fig. 9.10: Energy landscape from pinning centres and the activation energy required for flux creep.

such as dislocation creep. In Chapter 8, we came across a special equation called the Arrhenius equation that governs thermal activation processes like diffusion. Exactly the same equation governs flux creep. The higher the temperature and the smaller the activation energy barrier, the more likely it is for flux creep to occur. The energy barrier height depends on how strong the flux pinning is as well as how much Lorentz force is pushing on the flux line.

So what is the consequence of flux creep on the properties of the superconductor? Instead of there being an abrupt step-change in resistance at the point when the Lorentz force exceeds the pinning force, there is a much gentler transition. This means that we have to be a bit more careful about how we define the critical current density as there is not a fixed current at which the resistance switches on. What we do is to define the critical current as the current at which the voltage (or electric field) exceeds a certain threshold level—usually 1 microvolt per centimetre between the voltage contacts (see Fig. 9.11). Not surprisingly, flux creep is much more of a problem when we want to operate the superconductor at 77 K (liquid nitrogen temperature) rather than at 4 K (liquid helium temperature).

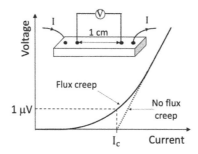

Fig. 9.11: Effect of flux creep on the current–voltage characteristics of a superconductor and definition of critical current using a 1 μV/cm electric field criterion.

Under the Lens

Dislocation creep

When materials scientists talk about *creep*, they are usually referring to a material gradually changing shape when a constant mechanical stress is applied. We have seen that plastic deformation happens because dislocations glide sideways on slip planes in response to a shear stress. When they encounter obstacles (other defects) on their glide planes they come to an abrupt stop. If the applied stress is increased, the dislocations may be able to cross over onto a different slip plane temporarily to get around the obstacle. This situation is not unlike a pinned flux line in a superconductor. Dislocations and flux lines are both linear features, and both are being trapped by interaction with some other defect in the material. We can persuade them both to move by brute force methods; flux lines will break free from the pinning defects if we increase the current to a point that the Lorentz force exceeds the pinning force, whereas trapped dislocations can move if the applied shear stress is increased.

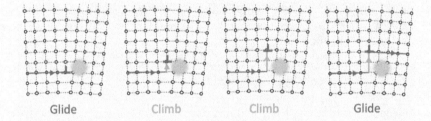

Fig. 9.12: Dislocation climbing to get around an obstacle.

Dislocation creep occurs when the stress is held constant and is insufficient to activate a second slip plane. This is very similar to flux creep in superconductors that happens when the Lorentz force is insufficient to overcome the pinning force. Both are thermally activated processes, so they happen more easily at higher temperatures. So how do the dislocations manage to move during creep? Remember, a dislocation can be thought of as the edge of an extra half-plane of atoms squeezed into the crystal. If vacancies (missing atoms) diffuse through the material to the dislocation, the extra half-plane loses a row of atoms from the edge and shrinks a bit. This so-called *climb* process moves the dislocation to an adjacent slip plane. If this happens repeatedly, the dislocation can gradually make its way to a plane that is not affected by the obstacle, and it can then glide sideways again without hindrance (Fig. 9.12). Since dislocation climb involves diffusion of vacancies, it happens much more easily at higher temperatures. Mechanical creep can actually happen via a number of different mechanisms, but most of them involve the diffusion of vacancies in some way or another.

There is another interesting consequence of flux creep in terms of the design of the right microstructure for optimising flux pinning. Within limits, it is generally the case that increasing the density of flux pinning sites will increase the pinning force per unit volume and will increase the critical current density. However, when the pinning sites get very close together, the activation barriers for moving flux lines between adjacent pinning sites decrease in height, making it easier for flux creep to occur. If flux creep is easier, the voltage criterion will be reached at lower currents. So, the optimum flux pinning landscape in high temperature superconductors will depend on exactly what temperature and magnetic field conditions you want to operate at. Materials optimised for high-temperature/low-field applications will be quite different to materials optimised for low-temperature/high-field. The first are likely to have a much lower density of flux pinning centres than the second.

9.7   Applications of high temperature superconductors

Now we have seen how it is possible to make brittle HTS materials in the form of long length wires or tapes capable of carrying enormous currents, what might we want to use them for? Leaving aside electronic device applications for the moment, there are two main classes of large scale applications: high-temperature/low-field applications, and low-temperature/high-field applications. The first category takes advantage of the lower cooling costs and simpler engineering associated with using liquid nitrogen as the coolant rather than liquid helium. The problem is that the current-carrying properties of HTS materials at the boiling point of liquid nitrogen (77 K) are not very good, especially when there is any appreciable magnetic field. This means that liquid nitrogen operation is limited to rather low field applications like power transmission cables. The second category of applications exploits the fact that HTS compounds can carry really impressive current densities even at very high magnetic fields—far beyond the field at which Nb_3Sn stops superconducting—provided they are cooled to temperatures well below their transition temperature.

Power transmission cables

In the developed world, it is taken for granted that we have all our electricity needs on tap in our homes. All we have to do is to flick a switch and the light comes on, or our kettle boils the water to make a cup of tea. We only have to think about what happens when we get unexpected power cuts to see how much of our everyday life depends on electricity. But how does the electricity get to where we need it, and how much of it is lost on its way? In the UK, all the different kinds of large electricity generators, from wind farms and solar farms to coal fired power stations to nuclear reactors are connected by a network of high voltage power lines (typically at 275 or 400 kilovolts (kV)) to substations in our local neighbourhoods. These long distance power lines are usually overhead—they are carried across the countryside by giant pylons. At the substations, the voltage is reduced to somewhere between 2 kV and 35 kV for distribution to our houses. In urban areas this distribution is usually done by cables buried underground, but in rural areas it is often overhead again. Finally, before the electricity enters our homes the voltage is dropped again to 240 V—what we call *mains electricity*—which is safer for domestic use. The reason that electricity is transported long distances at much higher voltages than we use in our homes is that the higher the voltage is, the less power is dissipated overcoming electrical resistance. Transmission at higher voltage is much more energy efficient. The voltage is stepped down to safer levels near our homes using transformers, and this technology relies on the electricity being carried as an *alternating current* (*AC*) that keeps changing direction backwards and forwards, rather than as a continuous *direct current* (*DC*). In the UK our grid operates at a frequency of 50 hertz, which means that the current oscillates 50 times every second.

So where do superconductors come in? As our global energy needs continue to grow, it is likely that we will need to upgrade the power grids to increase capacity. There is a general desire to do this by using more underground cables rather than building more unsightly overhead power lines that involve the destruction of large swathes of land-scape, even though digging the tunnels for underground cables is often significantly more expensive. Superconducting cables offer two main benefits over conventional underground cables. Firstly, a lot more current can be squeezed through a superconducting cable than a conventional (non-superconducting) cable of the same diameter. This means that it would be possible to upgrade the capacity of the network by replacing existing conventional underground power cables with new superconducting ones in the same tunnels. This retrofitting would be particularly beneficial in densely populated urban areas like Tokyo, where the cost of building new tunnels is prohibitively expensive. There are also advantages of using superconducting underground cables when overhead cables are being replaced because the size of the trench that needs to be dug for a superconducting cable is much smaller than for a conventional underground cable.

The second benefit is that superconducting cables potentially have lower losses than conventional ones. Now, why do I say *potentially*? Surely, superconductors carry electricity with zero resistance, so there should be a huge efficiency saving if we use superconductors rather than normal metals? Well, not necessarily. Firstly we need to remember that superconductors only work if we cool them to below their transition temperature and that will cost us some energy. For low magnetic field applications like power transmission cables, we can use liquid nitrogen to cool either Bi-2223 or (RE)BCO, and because this is such a good coolant, it will not use too much energy to keep the superconductor cold. A second reason, which may be less obvious, is that superconductors only actually have zero resistivity under DC conditions when the current flowing is not changing with time. Under AC conditions, every time the current changes direction, the magnetic field that it produces also changes direction. In wires capable of carrying high current densities, the magnetic flux lines that penetrate the superconductor must be strongly pinned, and this means that they cannot redistribute themselves as the field changes in the way that would minimise their energy. The upshot is that this strong pinning of the flux lines costs us a certain amount of energy each time the field is cycled. The more times we cycle the field back and forth per second, the more power is dissipated. This effect is called *AC loss* because it is the power loss associated with the alternating current.

So how can superconducting cables be designed to minimise the costs associated with

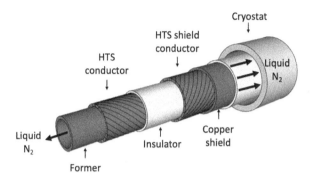

Fig. 9.13: HTS transmission cable with helical windings.

cooling and AC loss? The cooling part is relatively straightforward, typically involving flowing liquid nitrogen under pressure through a hole in centre of the cable and wrapping the HTS tapes that carry the current around this core. Thermal insulation is placed around the outside to minimise thermal losses to the environment. This design is not particularly innovative—it is rather similar to the design of liquid natural gas pipelines. The engineering considerations for minimising AC losses are less obvious. The first thing to realise is that the AC loss in the superconductor itself depends on the width of the conductor in the direction that is perpendicular to the magnetic field. This means that, in general, AC losses can be reduced by splitting the superconductor into many separate filaments—narrower wires—separated by non-superconducting material called the matrix (silver in the case of BSCCO multifilamentary wires). But there is another problem. If we just have a long straight wire, an electric field would be generated sideways between the filaments and this would lead to currents flowing in the matrix. The matrix is not superconducting so these currents will be subject to resistance leading to power dissipation. This effect of current being shared between the filaments by flowing through the matrix is referred to as *coupling* and it generates additional AC loss. The engineers' solution to reduce this unwanted coupling and minimise AC loss is to twist the wires. This works well for round wires, but for HTS tapes that are more difficult to twist, they are often wound in a loose spiral (helix) around a central cylindrical core. In fact, because there needs to be a return path for current as well in an AC cable to provide electromagnetic shielding,[8] typically there is an inner layer of tapes wound in one direction that carries the current to the load, and an outer layer wound in the opposite sense that carries the shielding current (Fig. 9.13). The counter-winding minimises coupling and reduces AC loss overall.

---

[8]This is just like the co-axial cables used for TV aerials which have one conductor down the middle and another in a tube around the outside.

There are several other factors that come into play in practice. For example, imagine we have done a really good job of stopping currents flowing from one filament or tape into an adjacent one. What would happen if a filament was damaged somewhere along its length and the supercurrent could no longer get through? What we would want is for current to easily be able to redistribute into the other filaments surrounding it without losing too much energy, but this is difficult if we have insulated our filaments from each other really well. As with most things, the best design is some sort of compromise between different factors. Another engineering decision that needs to be made is how tightly to wind the helix around the core. A very tight helix with a lot of turns along the length would generate large magnetic fields that limit the current that can be carried and would also use a huge length of conductor which would be expensive. But if we wind the conductor in a very loose helix, AC losses from the coupling would be high. In practice something in between is usually the right answer!

If we want to get electricity further afield—getting between countries for example—the most efficient way to do it is probably to use high voltage direct current (HVDC). This is because the longer the cable, the higher the AC losses will be. For relatively short distances the additional losses from AC transmission are offset by the cost of having to convert from AC to DC and back again—which is an expensive process. But for longer distance transport of electricity it becomes more cost effective to convert to DC and transmit with lower loss. It is particularly beneficial to use DC transmission cables when connecting the grid of one country to another because they will not necessarily be operating at exactly the same frequency and will not be synchronised with each other. These HVDC cables are known as *superhighways*, even though they conventionally use normal metal conductors rather than superconductors. Making superconducting superhighways is an attractive prospect because superconductors are ideally suited to DC operation where their zero resistivity property can be fully exploited. A striking way of comparing the efficiencies of conventional and superconducting technologies for long distance DC power transmission is by looking at the amount of carbon dioxide gas that would be released into the atmosphere just to feed the power that is lost in transmission. One study (Thomas *et al.*, 2016) has showed that for a 3000 km long 10 GW power line, the CO_2 emissions associated with the power lost from a conventional HVDC line would be equivalent to three standard coal fired power stations. By comparison, an HTS version, even taking into account the cooling power, would only produce the CO_2 equivalent to about 0.03 coal fired power stations—a whopping 100 times less CO_2!

We have seen that replacing conventional cables with the same diameter of superconducting cable can both increase capacity and improve efficiency, and cooling with liquid nitrogen is not a major challenge. So why does our power not zip across the countryside through super-efficient superconducting cables? The elephant in the room is that high temperature superconductor is eye-wateringly expensive. Although the cost of Bi-2223 and Bi-2212 is always going to be expensive because of the silver used as the sheath, (RE)BCO coated conductor costs are dominated by the cost of processing (rather than the raw materials) and so it is just conceivable that it will get cheaper as manufacture is scaled up. What may help is a major large scale application that absolutely has to have high temperature superconductors because it will not work without. This would allow the coated conductor manufacturers to scale up production and benefit from economies of scale. If the costs can be driven down to somewhere in the region of 50 dollars per kiloamp metre (that is 50 dollars for every metre of conductor that can carry 1000 A of current) then HTS materials may become economically viable for many more applications including transmission cables and other power applications like motors and generators.

High field magnets

The other fabulous property of HTS materials is that they can carry enormous currents in very high magnetic fields—much higher fields than even the best low temperature superconducting (LTS) material (Nb_3Sn)—so long as we cool them down to a low enough temperature. For example, (RE)BCO can still work in magnetic fields well in excess of 20 tesla even at the relatively high temperature of 20 K. (Remember that one of those really strong permanent magnets that you can get hold of can only produce a magnetic field of about 1 tesla, so 20 tesla is a very strong field.) But what do we need such high magnetic fields for? One simple answer is that researchers always crave higher and higher magnetic fields for testing the properties of other materials. For example, nuclear magnetic resonance (NMR) instruments are very commonly used to identify the structures of molecules such as those used in new drugs. Like its sister technique MRI (discussed in Section 2.2), NMR relies on a strong background field. The stronger the field, the more sensitive the instrument is. This means that there is always a drive to make higher magnetic field instruments that are capable of doing new science. NMR is a big market, but there are other challenges to overcome if we want to use HTS materials in this application. One of these is that, just like MRI, NMR magnets need to be incredibly stable—the magnetic field must stay absolutely constant with time. The best way to achieve this in practice is by running the magnets in persistent mode, without a power supply. This relies on the magnets being a continuous superconducting circuit, joined together using ultra-low resistance superconducting joints. We have had decent ways of doing this for low temperature superconductors for a long time, but it is much more difficult to make truly superconducting joints between high temperature superconductors. Even if we could make good contacts between the oxide HTS materials using the same kind of superconducting solders that we use for joining LTS wires, these solders cannot withstand high magnetic fields so the joints would have to be positioned a long way outside the magnet windings. Also, these ultra high field magnets typically use NbTi for the outer windings (because it is nice and cheap), with Nb_3Sn inside that where the field is higher, and HTS to boost the field right in the middle of the magnet. This reduces the length of the expensive HTS conductor that is required, but it makes it difficult to locate joints in low enough fields for superconducting solders to work. Other methods for joining HTS materials are also tricky because high temperatures must be avoided in the process because we do not want to mess up the carefully optimised microstructure of the superconductor.

On top of this, to get the magnetic field quality required for NMR, it is also desirable to connect the different kinds of superconducting wires in one long length so that the outer and inner magnets are in series with each other rather than in parallel.

This means that ideally we also need to make superconducting joints between Nb_3Sn metallic superconductor and ceramic HTS compounds. This is a big problem from a materials science point of view. One reason is that metals have the tendency of wanting to pick up oxygen if they can, and HTS materials have a nasty habit of losing their oxygen and with it their superconducting properties. Even if we can find a way of preventing the very fussy high temperature superconductor directly reacting with the LTS metal alloy, the metal is likely to suck oxygen out of the surface of the HTS material and this will completely muck up the superconducting properties of the surface layer in the joint area leading to unwanted resistance.

Ultra high field research magnets—ones that are used for testing the fundamental physical properties of materials—do not need truly superconducting joints because they do not need the same level of stability as an NMR magnet. The world record for an all-superconducting magnet at the time of writing is 31 tesla at the National High Magnetic Field Laboratory in Florida. This uses an LTS magnet with a big hole in the middle that a separate (RE)BCO HTS magnet is placed inside. Although these magnets are remarkable feats of engineering, they are one-offs that are located at specialist national facilities, so are not likely to be the mass-produced items that are needed to drive down coated conductor manufacturing costs. The same is true for NMR magnets. The highest field, state-of-the-art instruments will only be affordable for the big research labs, with smaller and cheaper instruments being the mass-produced item.

What we are looking for are major large scale applications that are enabled by HTS technology and will require huge lengths of conductor so that manufacturing can be scaled up and (hopefully) costs can be driven down. Two future applications that might fit the bill are discussed in detail in Chapter 10: large particle accelerators for smashing atoms together, and nuclear fusion reactors for energy generation. For both of these applications, increasing magnetic field has large paybacks, so there are good arguments for using high temperature superconductors. The conductors will need to carry enormous currents, somewhere in the order of tens of kiloamps (10,000–70,000 amps). Individual (RE)BCO tapes or BSCCO wires cannot carry anywhere near this kind of current, so it will be necessary to make cables consisting of many strands. Design of these cables is complex, particularly for accelerator magnets that need to have very good field quality.

(RE)BCO bulks

In Chapter 4 we saw that if we have pellets of superconductor—known as *bulks*—we can get them to levitate or we can turn them into incredibly strong and compact permanent magnets for desktop NMR/MRI or for electrical machines. These applications rely on setting up circulating currents in the pellets, and the higher the critical current density of the superconductor the better. We have already seen that to get high critical current densities in HTS materials we need to do three things: put in lots of flux pinning defects, eliminate high angle grain boundaries that block current flow, and make sure the chemistry is just right so that T_c is not compromised. (RE)BCO is the superconductor of choice for these bulk applications because it has decent critical current density at 77 K so we can use liquid nitrogen as the coolant.

The superconducting (RE)BCO phase, RE-123, is made in bulk form by melting it and letting it re-solidify in a process rather similar to the one described for Bi-2212 wires. We start by packing RE-123 powder into a cylindrical mould and squashing it so that it loosely binds together into a pellet. When it is heated up to somewhere around 1000°C it partially melts leaving particles of a non-superconducting solid phase dispersed in a liquid. This solid phase is called RE-211, and it contains the same constituent elements as the superconducting phase, but in a different ratio. If we cool the partially melted mixture back down slowly enough, the reverse reaction happens and the superconducting RE-123 compound reforms by reaction of the RE-211 with the liquid.[9] The problem is that, unless we use a cunning trick, the RE-123 grains that form from the melt will be aligned in random directions and so there would be lots of grain boundaries. A small single crystal of something with a higher melting point is placed on the top of the pressed pellet before it is melted. We call this the *seed* because when the desired RE-123 phase reforms on cooling it starts to grow outwards from the seed, copying its crystal orientation. The seed crystal acts as the template in the same way that the substrate acts as the template in coated conductors. Provided the cooling rate is really slow, it is possible to stop crystals forming from anywhere else and we end up with a large single grain of (RE)BCO. This is exactly the same principle that is used to grow large single crystals from a saturated solution. You may have done this yourself with something like sugar crystals by tying a small sugar crystal on a piece of thread and suspending it into a saturated sugar solution to seed the growth of a large sugar crystal. We call the bulk pellets that are formed single grains, although they actually grow outwards from the seed in five directions simultaneously, making five different growth sectors that join together, as shown in Fig. 9.14. All of the growth sectors have the same crystal orientation because they all nucleate from the same seed, so the boundaries between the growth sectors are not the same as grain boundaries which separate misoriented crystals. That being said, the crystallographic alignment is far from perfect in these 'single grain' bulks. Each growth sector contains a mosaic of subgrains that are slightly misoriented from each other. They are elongated along the growth direction and can even be seen by eye on the top surface of a (RE)BCO bulk (Fig. 9.14(b)).

So how do we incorporate flux pinning centres to push up the critical current density? Well, one way is to be clever about the process and use the fact that when we melt the (RE)BCO pellet we naturally form particles of RE-211 that do not superconduct. If a bit of extra RE-211 is added to the starting powder and the cooling rate is just right, it is possible to make sure that some of these particles get trapped in the RE-123 crystal as it grows. This is a balancing act because you want to manage to grow a single crystal which requires very slow cooling, but also trap as much RE-211 as possible as you go which is easier with faster cooling (9.14(d)).

---

[9]This is another example of a *peritectic solidification* reaction, just like the solidification of Nb$_3$Sn and Bi-2212, which both occur by reaction of a solid phase with liquid.

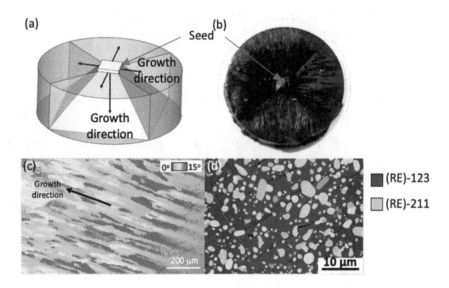

Fig. 9.14: Bulk (RE)BCO pellets. (a) Diagram showing seed and growth sectors. (b) Photograph of surface (reproduced from (Namburi *et al.*, 2021) under the Creative Commons Attribution 4.0 license http://creativecommons.org/licenses/by/4.0/). (c) Map showing misoriented subgrains taken using the electron backscatter diffraction technique in a scanning electron microscope. (d) SEM image of Y-211 particles in a Y-123 single grain sample (courtesy of Dr. T. Mousavi, University of Oxford).

Finally, we have to address the chemical composition. After the melt-process, the solid (RE)BCO pellets do not have enough oxygen in them so they do not superconduct as well as they ought to. We get them back to the sweet spot by giving them a cook in an oxygen atmosphere at a temperature that is not high enough for them to melt again, but allows oxygen to make its way all the way back into the centre. This takes the best part of a week, even though the bulks are full of holes and cracks! It's worth the wait because the final pellets have high T_c, no high angle grain boundaries and are riddled with small RE-211 particles that act as flux pinning sites—the ideal microstructure for getting high current densities. In fact, the current density in the best bulk samples is sufficiently high that their performance is limited by when they physically crack apart under the magnetic forces rather than because they stop superconducting. The world record bulk magnets are actually put into compression before charging them up by shrink wrapping them in stainless steel collars to give them a bit of extra mechanical performance.

The Wider View

Single crystal turbine blades

Turbine blades in jet engines—the ones used to power aeroplanes—are exposed to an incredibly harsh environment. The engines operate at as high a temperature as possible for maximum efficiency, but this means that some of the turbine blades have to operate at about 1700°C—several hundred degrees above the melting point of the nickel alloy that they are made from! To stop them from melting, the blades contain a network of complex cooling channels for gas to flow through, and the outside is coated with a ceramic to provide further protection. This brings the temperature of the blades down to below their melting point, but they are still operating at a very high proportion of their melting point. The blades are rotating at very high speed—about 12,000 rotations per minute—and that brings with it high mechanical stresses. This combination of large stresses and very high temperatures is a recipe for creep. The tolerances are very tight in the engine though, so there is limited space for the materials to deform without the moving parts touching.

Superalloys are special materials that have been designed for operation at very high temperatures close to their melting points. The archetypal superalloy for aeroengines is a nickel-based alloy with the main alloying element being aluminium (or titanium). It gets its extraordinary properties as a result of its special two phase microstructure. The main phase is known as 'gamma' (γ) and it consists of a face-centred cubic structure with some of the nickel atoms replaced at random with aluminium atoms. Within this matrix are precipitates of a second phase called 'gamma prime' (γ'). The crystal structure is almost identical to γ, but instead of the aluminium atoms occupying random positions, they occupy the corner sites in the unit cell, with the nickel atoms occupying the face-centres. This is known as ordering and we came across it before when we talked about intermetallics like Nb_3Sn. Dislocations find it very difficult to move through ordered structures, because their movement would disrupt the ordering, which is why Nb_3Sn is brittle. In the case of a nickel superalloy, the presence of the γ' precipitates provides mechanical strengthening. There is another reason why this γ/γ' microstructure is particularly effective though, and it comes down to the fact that their crystal structures are so similar that the precipitates line up in the same orientation as the matrix. But why does this make a difference? The key thing is that, because the atomic planes are continuous at the boundaries, the energy cost associated with creating the interface is very low. Since it is interface energy that usually drives the coarsening of precipitates (because larger particles have a smaller surface area to volume ratio), the low interface energy in superalloys means that there is little driving force for growth and the γ' particles stay small. This makes them much more effective at strengthening the material for service at high temperature for long periods.

Creep will still be a problem though, and it turns out that grain boundaries are one of the main culprits. In addition to the dislocation glide-climb mechanism of creep discussed earlier, at very high temperatures grains can slide sideways over one another in response to an applied force, leading to deformation. The best way to avoid this happening is to get rid of the grain boundaries altogether—make the sample as a single grain. Turbine blades made this way are usually called *single crystal* turbine blades, even though they contain two phases. This is not unlike the microstructure we get in single grain (RE)BCO bulks in which the matrix is pretty much a single crystal, but it contains precipitates of a different phase. Single crystal turbine blades were initially developed in the 1980s, and they are made by casting the alloy from the melt in a special way by withdrawing the mould from the hot zone of the furnace gradually to encourage directional solidification from one end (and avoid nucleation of solid anywhere else). The first bit of solid that forms is not single crystalline, but as the crystal is withdrawn the grains get elongated along the growth direction. Then comes the clever bit. The mould has a special spiral shape, called the 'pigtail' that the grains have to pass through to get into the main bit of the casting. Only one of the grains manages to get through this spiral, and the rest of the blade is grown from this single grain (figure 9.15).

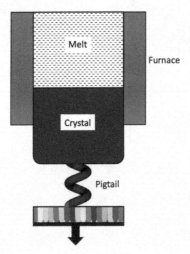

Fig. 9.15: Apparatus for growing a single crystal superalloy turbine blade.

9.8   Small scale stuff

Large scale applications of high temperature superconductors have yet to become properly commercial because of the complexities of the materials processing and the cost of buying long lengths of wire. However, we must not forget that superconductors can be used for a wide variety of small scale devices as well. These are typically divided into two categories: passive devices and active devices. Active devices tend to involve switching between two states by application of a signal and usually work using the Josephson junction technology described in Section 6.4. We can make HTS versions of many of the standard LTS devices, including HTS SQUIDs that can be flown over the landscape in aeroplanes and can pick up magnetic fields from metal ore deposits underground. On the other hand, passive devices do not rely on Josephson junctions. Instead they are special AC circuits that interact with electromagnetic waves. Our long distance communications systems typically use radio waves or microwaves which have much longer wavelengths than visible light. Mobile communications, like mobile phones, typically use microwave frequencies with wavelengths of the order of a few centimetres, and one of the main challenges with our increasing use of mobile communications is that the frequency space for carrying signals is getting very crowded. Basically, the spectrum is divided into a series of channels each corresponding to a particular band of frequencies. These channels are getting narrower and narrower in frequency to cope with increased demand. The problem is that we need to make sure that there is no interference between the signals from adjacent channels. To do this we use devices called filters. A *bandpass* filter is a special circuit that transmits a specific frequency band of signals but rejects (does not transmit) any other frequencies. We need these kinds of devices on the front-ends of the receiving base stations for our mobile phones so that they can sort out which signals to 'listen' to and re-transmit.

So, how do they work? If you imagine a perfect pendulum with no friction and no air resistance, regardless of how high we lift it to start it swinging, it will always take the same amount of time to get back to the place where it started. The number of times it goes back and forth per second—its frequency—will be a constant unless we change the length of the string it is suspended on. We call this frequency its natural frequency—the frequency it naturally wants to oscillate backwards and forwards with. The same is true for a swing. Depending on how long the chains suspending it are there is a natural frequency that it will swing at. If we push the child at exactly the right time, they will swing higher and higher, but if we push them at the wrong time they will instantly lose height. So what has this to do with mobile phone filters? Well, the circuits that make up filters are just like pendulums and swings. If you drive them at the right frequency—their natural frequency—you can increase the amplitude of the signal, but if you drive them at the wrong frequency the signal will be cut dead. This effect is called resonance, and it is exactly the same physics that is exploited in magnetic resonance imaging discussed in Section 2.2. To make a bandpass filter we put together a series of circuit elements each of which resonates at a slightly different frequency. Together they will accept a chosen range of frequencies and reject all the others.

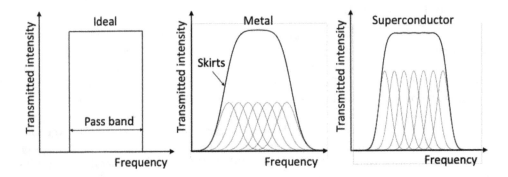

Fig. 9.16: Bandpass filter characteristics. (a) Ideal top-hat shape, (b) conventional metal resonator filter with wide skirts, (c) high temperature superconducting resonators with narrower skirts.

OK, but why superconductors? Ideally, if we plot the transmitted intensity as a function of frequency for a bandpass filter, we want it to have perfectly vertical sides and a nice flat top in the pass band (as shown in Fig. 9.16). However, conventional resonator circuits tend to have rather wide resonant responses because of the resistance in the circuit. Resistance essentially introduces the same effect as friction does in a pendulum. It damps the oscillation so that it loses height quickly. For a resonant circuit this means that the peak gets broader and less high. When we put a set of resonators of different overlapping frequencies together, resistance in the circuit means the edges of the passband—which we call the *skirts*—are rather more sloped than we want them to be. Now we can start to see why using superconductors with their lower resistance might be beneficial for these devices. However, as we have discussed, superconductors do not actually have zero resistance under alternating current conditions. In this low field case, it comes down to the fact that in a superconductor there are normal unpaired electrons that are free to move as well as superconducting charge carriers (Cooper pairs). Under direct current (DC) conditions, the current will be carried exclusively by the Cooper pairs because they can travel without resistance. However, under alternating current (AC) conditions, because the electrons have some inertia—they find it difficult to change direction quickly—the supercurrent will tend to lag a bit. The upshot of this is that the normal (non-superconducting) electrons end up carrying a bit of the current. However, despite this, high temperature superconductors still have considerably lower resistance than copper at microwave frequencies. This means that the resonance is sharper, and filters with steeper *skirts* can be produced in much smaller devices.

These superconducting filters are typically made using thin films on single crystal substrates about 5 cm in diameter, and so they are relatively easy components to keep at liquid nitrogen temperature with a little cryocooler in a base station. In fact, in the USA there are over 10,000 base stations that use HTS filters. The small size of these filters also makes them very attractive for satellite communications, as every gram of weight that can be saved is valuable if you have got to get it launched into space, and there have already been several experiments including the worldwide High Temperature Superconductivity Space Experiment, in which an HTS filter orbited earth for 18 months. HTS filters have also been used in radio astronomy applications where very small signals often suffer from electromagnetic interference from television signals for example.

Chapter summary

- Although high temperature superconductors have taken longer to take off than initially predicted when they were first discovered, truly remarkable materials science has enabled us to get these troublesome ceramics into a form that can be useful in real applications.

- They are incredibly challenging materials to work with for four main reasons: extreme chemical sensitivity, the grain boundary problem, anisotropic properties, mechanical fragility.

- In spite of the immense challenges of working with these intractable materials, phenomenal performance has been achieved but at a cost that is simply too expensive for them to break into the commercial market.

- HTS materials are already commercially used in some mobile communications systems and other small scale devices like SQUIDs.

10

A Super Future?

The initial hype surrounding the discovery of high temperature superconductors was largely based on the simple fact that using liquid nitrogen instead of liquid helium is much cheaper, easier and more socially acceptable. Well that, and the captivating idea that it might herald the arrival of the holy grail—a room temperature superconductor that would not need cooling at all! As we have seen in Chapter 9, things did not turn out to be as simple as all that. For one thing, it proved to be difficult to realise the high currents needed to make large scale applications a reality. Furthermore, although many cuprate compounds with higher critical temperatures were found in the few years following the initial discovery—the best one being a mercury-based compound with a T_c value over 130 K—despite a great deal of effort, the field then ground to a halt and no higher T_c compounds were discovered. Even more disappointingly, it has proved very difficult to understand how high temperature superconductivity actually works, so although there are many theories, we are not much further forward with knowing where to look for brand new high temperature superconductors.

The last 30 years have not been wasted though. Many of the technical challenges have been overcome with very clever materials science, but the materials are still prohibitively expensive for most applications. What is needed is a really important large-scale application that simply cannot work without high temperature superconducting technology. This could provide the market pull that is needed to scale up conductor production and bring down cost, as well as drive further materials developments. If that happens, lots of the other technologies would become commercially viable. So what might that revolutionary application be? One possibility is the *Future Circular Collider*—the big brother of the Large Hadron Collider—which is still at a conceptual stage but will need higher field magnets than the LHC. However, this is a single machine, albeit a very large one that will need an enormous amount of superconductor, and it may end up being made with Nb3Sn rather than an HTS material. Another possible application is commercial nuclear fusion energy generators which need very high magnetic fields to confine the intensely hot plasma—hotter than the sun—where the fusion reaction takes place. It is interesting to note that in both of these applications, it is the extremely good high field properties of HTS materials that makes them essential, rather than their ability to operate at higher temperatures, although as we will see, it is likely that fusion magnets will be operated at around 20 K rather than 4 K. But before we come on to discussing these future applications for existing superconducting materials, let's explore how likely is it that we will discover that holy grail—a room temperature superconductor—and if we do, would it really be revolutionary?

10.1   Room temperature superconductivity

I think it is worth starting this section by coming clean with you; we already have a room temperature superconductor! In fact, room temperature superconductivity has been convincingly demonstrated in two different kinds of experiment. The first of these was an experiment in 2013 in which room temperature superconductivity was observed in YBCO (a high temperature superconductor) using a short infrared laser pulse. The problem was that room temperature superconductivity only lasted for a few millionths of a microsecond. The second major breakthrough happened in 2015 when supercon-ductivity was found at 203 K ($-70°C$) in a hydrogen sulphide compound (Drozdov *et al.*, 2015). In the couple of years that followed this discovery, very high temper-ature superconductivity has been predicted theoretically in lots of different hydride materials, and others have been found experimentally. Notably, superconductivity at somewhere between 250 and 260 K (about$-20°C$) was found in a lanthanum hydride with stoichiometry somewhere around LaH_{10}. These metal hydrides containing lots of hydrogen are referred to as *superhydrides*. In late 2020, this series of discoveries was followed up by a report in one of the most famous scientific journals, *Nature*, of su-perconductivity in a 'carbonaceous sulphur hydride' at 288 K ($15°C$) which can safely claim to be 'room temperature' most of the year round if you live somewhere with a temperate climate like the UK (Snider *et al.*, 2020)! Certainly, since our household fridges are at about $4°C$ we certainly do not have any concerns about cooling these superconductors.

So will these hydrogen compounds be revolutionary? Well, I am sorry to say that, no, they are very unlikely to be useful technologically. The reason is that, although they work at or close to room temperature, they only do so under unimaginably high pressures. The pressures that are needed are upwards of 150 gigapascals—a whopping 1.5 million times higher than atmospheric pressure. It is almost half the pressure at the centre of the earth! These kinds of pressures are much more difficult to realise in practice than cooling down to 4 K using liquid helium, so the idea of using the superhydrides to make a large scale device is totally unrealistic. However, that is not to say that the discoveries are not truly remarkable, and I am sure they will provide new insights into the phenomenon of superconductivity.

10.2   Discovering superconductors

If we want to get some idea about how likely it is that we will find a more practical room temperature superconductor, it is worth reflecting on how new superconductors have been discovered in the past. The initial discovery of superconductivity in 1911 was a complete surprise. Nobody had predicted the phenomenon theoretically or had any inkling that a material might behave that way. In the decades that followed, lots of new superconductors were discovered by trying out different materials, replacing one element with another and seeing what happened. People started to get better at figuring out what might make a good superconductor. Then, in 1957 there was a breakthrough with understanding what was going on at the microscopic level in these superconductors when Bardeen, Cooper and Schrieffer came up with their quantum theory for superconductivity which has become known as the *BCS* theory. This explained superconductivity by the linking of electrons into Cooper pairs that can travel through the material without being subjected to the kind of scattering that produces resistance in normal metals. They also discovered that the origin of the attractive force that holds the electrons together in a Cooper pair is vibrations of the atomic lattice—phonons.

At about the same time, a scientist called Bernd Matthias came up with a set of empirical rules based on his practical experience detailing where to look in the periodic table of elements if you want to find a new superconductor. His rules can be summarised as follows:

1. Crystals with high symmetry are good—cubic crystals are best.

2. There should be between two and eight valence electrons (outer shell electrons) per atom in the compound.

3. Stay away from oxygen.

4. Stay away from magnetism.

5. Stay away from insulators.

6. Stay away from theorists.

These rules were very successful, up to the point where high temperature superconductors were discovered. Let's take each in turn. The first says that we ideally want to have cubic crystals, but the cuprate HTS materials are highly anisotropic because of their layered structures—far less symmetrical than cubic crystals. Secondly, Matthias's rule for how many electrons should be in the outer shell completely fails for copper which, in its elemental state has ten electrons in its outer shell.[1] This does not fit with Matthias's second rule. Now, let's take the next three rules together. Cuprates are oxides and their parent compounds[2] are antiferromagnetic insulators. Many scientists consider that the only one of Matthias's rules that has not been 'disproved' by high temperature superconductors is the last one! That may sound harsh on theorists, but at the time of writing, we still do not have a definitive theory for why high temperature superconductors like the cuprates work.

---

[1] For those of you who are familiar with the periodic table, you would be forgiven for thinking that copper would have nine 3d electrons and two 4s electrons, but actually it saves itself energy to have the $3d^{10}4s^1$ full shell configuration.

[2] These are compounds before we dope them by taking out some oxygen for example to make them superconduct.

The discovery of high temperature superconductivity probably happened more by chance and intuition than anything else. By the mid 1970s a plateau in the maximum T_c had been reached at about 23 K, and try as they might, nobody seemed to be able to get anything higher. Superconductivity at 13 K had already been discovered in a couple of oxides, but their critical temperatures were not high enough to cause much interest. In the early 1980s, two researchers Alex Muller and Johannes Bednorz teamed up and focused their efforts on searching for higher temperature superconductivity amongst the oxides. They spotted a research paper by another group on a barium-doped lanthanum cuprate that behaved like a metal down to about $-100°C$ and recognised it as a sign that this compound might have interesting electrical properties at lower temperature. They discovered that if you kept cooling it down below $-100°C$, its resistivity went up again, before suddenly becoming superconducting at 11 K. This initial result was nothing to write home about, but just a short while later they had managed to play around with the composition and raised the T_c to 30 K—higher than had ever been found before. However, the field had been plagued with false claims in the past, so Bednorz and Muller were rather tentative when they first reported their result in a paper entitled 'Possible High T_c Superconductivity in the Ba-La-Cu-O system' (Bednorz and Müller, 1986). Their reservations stemmed from the fact that, although they had measured zero resistivity, they had not demonstrated the Meissner effect at that time. Later the same year it was confirmed, and it immediately attracted a huge worldwide effort to find more compounds along the same lines. It was a matter of a few months only before the scientists worked out that replacing lanthanum with another rare earth element, yttrium, produced the compound now known as simply YBCO, with a critical temperature of 90 K, and the rest is history.

So although the discovery of HTS certainly involved some luck, Bednorz and Muller were making a conscious effort to look for superconductivity in this kind of compound. The same was true for the discovery of magnesium diboride in 2001. It is rather a remarkable story, not so much because superconductivity was discovered in this compound, but because, unlike in the HTS compounds, it turns out to be a pretty conventional superconductor that follows BCS theory. So how come nobody had spotted it before? If we go back to Matthias's rules, we perhaps would not expect MgB_2 to be a good conventional superconductor because it neither has cubic symmetry nor does it contain a transition metal with between two and eight d electrons in its outer shell. However, it does contain light elements which we would expect should be good in terms of the phonons that are needed for forming Cooper pairs in conventional superconductors. It is also a very simple binary compound that had been manufactured in mass quantities for decades and does not need to be doped to make it superconduct. In fact, it has the highest transition temperature of 39–40 K when it is perfectly stoichiometric with exactly two boron atoms for every magnesium atom. Moreover, as mentioned in Section 3.6, data showing an anomaly in heat capacity—a signature of a superconducting transition—had already been published in 1957 (Swift and White, 1957). Jan Akimitsu's group were actually looking for superconductivity in much more complicated compounds involving magnesium, boron and a transition metal like titanium when they chanced upon the unexpected discovery of superconductivity in the simpler MgB_2 compound (Nagamatsu *et al.*, 2001).

The Wider View

Transparent conducting oxides

Transparent conducting oxides are a rather important class of materials because without them touch screen devices would not work. It is actually pretty difficult to find materials that are both conducting and transparent. Think about the good electrical conductors that you know. What do they look like? Well, my guess is that you thought of a chunk of metal, and metals are typically optically opaque and shiny. Not transparent. Now think about transparent materials—glass perhaps. They are not known for being good electrical conductors! We can understand why, if we think about the physics. Something is transparent if light can get through it without interacting very much. Light is just an electromagnetic wave, so it has an oscillating electric field. When light is incident on a conducting material like a metal, which is full of free electrons, those electrons should oscillate backwards and forwards, being pushed and pulled by the constantly changing electric field of the light wave. This absorbs energy from the light and so it does not pass through the material, instead being re-emitted backwards by the oscillating electrons and making it look shiny. Indium tin oxide (ITO) is the transparent conducting material that is used pretty much universally at the moment. It sits somewhere at the sweet spot, having enough free electrons to be conducting without having so many that all of the light is absorbed. The reliance of the technology on indium is a problem though, because it is scarce and expensive, so there has been a major research effort to look for suitable alternatives. It was a team in China led by Hideo Hosono who were looking for new transparent conducting oxides that happened upon the iron-based superconductor.

The discovery of high temperature superconductivity in iron-based compounds in 2006 was even more of a surprise (Kamihara *et al.*, 2006). The team were not even working on superconductors. Instead, they were looking for materials that would be good transparent conducting oxides. In fact, in recent times, it is only the hydride superconductors that were predicted to be high temperature superconductors before they were measured experimentally. The idea that if you could get hydrogen at high enough pressures to make it a metallic solid it would be a conventional BCS superconductor at high temperatures was predicted as long ago as 1968 by a scientist called Neil Ashcroft. Although nobody has managed to make metallic hydrogen yet, the hydride superconductors do seem to be conventional BCS superconductors, unlike the unconventional and unexplained type of superconductivity in the cuprates and iron-based materials.

So where do we go from here in terms of looking for new high temperature superconductors that work at ambient pressures like the cuprates? Well, the problem is that the periodic table contains something like 118 elements, 94 of which occur naturally on Earth and our experience of high temperature superconductors is that the best ones contain four or five different elements at least. This would make at least 3 million different possible combinations! On top of that, we know that we would probably have to get the ratio of the different elements spot on and we may well find that even if we make the correct compound, it does not superconduct until we tweak the chemistry in some way to get the doping right. It is simply like looking for a needle in a haystack unless we have some practical search strategies. The problem is that we do not know why unconventional superconductors work, and until we have a more definitive idea of what the glue is that holds the Cooper pairs together, it is very difficult to work out sensible rules for the chemists to use to find new compounds. Once we have hit upon a family of compounds, like the cuprates or the iron-based materials, the chemists are amazing at playing lots of clever tricks like replacing phosphorus with arsenic, or replacing lanthanum with yttrium to find similar compounds that work better, but it is very difficult to predict from first principles which compounds might be high T_c superconductors because we simply do not have sufficient theoretical understanding. That does not mean that we will not find any better superconductors or that they do not exist—in fact, I would be very surprised if we had already managed to chance upon the best the periodic table has to offer—but when we will find them is pretty much anyone's guess!

10.3   Novel superconductors

So far this book has concentrated almost exclusively on the technological supercon-ductors—the six superconductors that you can go out and buy commercially in the form of wires that carry lots of current. Of course, there are a huge number of dif-ferent superconducting compounds—far too many to do them all justice in a single book! Leaving aside the conventional metals and alloys that can be explained by the BCS theory, there are plenty of other types of unconventional superconductor that are fascinating, either because they display novel physics that might be exploitable in the future or because they give us a hint of the mechanisms responsible for unconven-tional superconductivity. These include the organic superconductors and iron-based superconductors that we will talk about in a bit more detail here, but also exotic new materials like 'topological superconductors' which are predicted to support a special particle called a Majorana fermion that people think will be useful in the future for fault-tolerant quantum computing.

Materials by design—the organic superconductors

Carbon—the basis of organic compounds—is an incredibly versatile atom because each one can form up to four covalent bonds either with other carbon atoms or with atoms of other elements—most notably hydrogen, nitrogen, oxygen and phosphorus. It can form small molecules like methane (CH_4) or carbon-dioxide (CO_2), or it can form long chain polymer molecules (which we often refer to generically as plastic), or it can form giant covalently bonded structures like graphite or diamond. It can be stable as a single two-dimensional sheet of hexagonally bonded atoms known as *graphene* as well as in the form of nanomaterials with other dimensionalities. Here a nanomaterial refers to at least one of its physical dimensions being below a 100 nanometres (10^{-7} m or 0.0001 mm) in length. The dimensionality of a nanomaterial typically refers to how many of its dimensions are on the macroscale (i.e. bigger than nanoscale). A 2D nanomaterial

Under the Lens

Computational materials discovery

One of the recent developments in computation that might prove to be helpful is the brute force approach to looking for materials that are similar to compounds that are known to be good superconductors. Essentially these methods involve matching some key signature of the good superconductor with the signatures of a vast number of other crystals in a huge database and seeing which ones give the closest match. But what characteristics of the good superconductor should be chosen to do the matching? Well, superconductivity, like many materials properties, ultimately comes down to the electronic structure—in particular the nature and occupation of the electron states near the Fermi energy.

Now, in cuprate high T_c superconductors, the electronic structure is pretty complicated. It is generally believed that superconductivity arises from the copper-oxygen planes in the crystal, and within these planes, the copper atoms form bonds with surrounding oxygen atoms. Copper is a transition metal which, in the metallic state, has completely filled electron 3d orbitals. There are in fact five atomic d orbitals, all of which have different shapes. Each d orbital can hold two electrons, making a total of ten electrons in a filled d-subshell. A simple picture of what happens in the copper-oxygen planes of high T_c cuprates is that one particular d orbital is involved in the covalent bonding with the p orbitals of the neighbouring oxygen atoms because it has lobes of electron density pointing in the right directions (see Fig. 10.1). The copper d orbital joins up with the oxygen p orbitals and forms a new orbital which is a hybrid of the two. In high T_c cuprates, this particular hybrid orbital (or band) has holes—missing electrons—which is known to be crucial for superconductivity. Some recent studies have used this special feature of the cuprate electronic structure as a signature to match against when data mining to look for other compounds that might have similar properties (Geilhufe *et al.*, 2018).

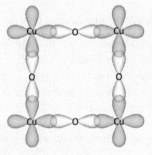

Fig. 10.1: Bonding between Cu 3d and O 2p orbitals.

This brute force data mining approach to searching for new superconductors gets around the embarrassment that we do not actually know the underlying mechanism for superconductivity by matching to special characteristics of the compounds that seem to be important. It is rather blinkered though, in so much as it will not find radically new types of superconductor where the underlying physics is different. We could take the iron-based superconductors as an example. If we use a search strategy based on their being one partially occupied d orbital (like the cuprates), none of the iron-based superconductors would match our search because they have several partially occupied d orbitals!

is a sheet with two macroscopic dimensions and one (the thickness of the sheet) being nanoscale. A 1D nanomaterial is shaped like a rod, with its length being macroscopic but the other two dimensions being nanoscale. A 0D nanomaterial is something that has all three dimensions being small enough to meet the 'nano' criterion.

Organic molecules, at first sight, may not seem to be likely candidates for superconductivity, not least because they are generally insulators and superconductivity tends to emerge from a metallic state of some sort. However, in the 1970s there was great excitement when superconductivity was discovered in organic 'metals'. These compounds tend to have very complicated names, so are generally referred to by acronyms. The most famous family are based on a molecule which has the catchy name 'BEDT-TTF' or 'ET' for short. These molecules are roughly flat, and can be stacked in a range of different ways to form a three-dimensional crystal. In the stacked structures the 'ET' electron orbitals—the regions in space where the electrons most like to reside—overlap with each other and allow electrons to move from molecule to molecule across the structure. In these periodic structures, the overlapping electron orbitals are typically referred to as *bands*. The stacking together of the 'ET' building blocks is achieved in practice by mixing them with molecules of something else—let's call them 'X'—that like to steal electrons from the 'ET' molecules. This electron transfer binds the molecules together in the same kind of way that an ionic bond works in inorganic materials. The result is that, because the 'ET' molecules have collectively lost electrons by donating them to the 'X' molecules, their electron bands are not completely filled up with electrons anymore. We call the empty states where the electrons have been removed *holes* and they are effectively positively charged because they are places where there should be a negatively charged electron but it has been taken away. These holes are free to move because electrons from neighbouring sites can jump into them one after another, and because the electron orbitals from adjacent 'ET' molecules have joined together into delocalised bands, the holes can migrate across the entire crystal. This ability for holes to move through the material means that we would expect it to conduct electricity. Metals conduct electricity in the same way because they also have partially filled delocalised bands—often referred to as the 'sea of delocalised electrons'.

Interestingly, the flat nature of the 'ET' molecules and their relations means that the electronic properties of the crystals derived from them are rather anisotropic (different in different directions). In the 10 K class compounds, the 'ET' molecules stack them-

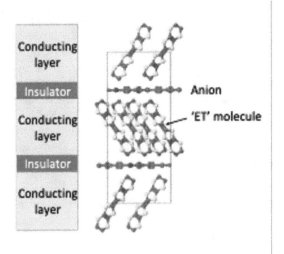

Fig. 10.2: $(ET)_2$-$Cu[N(CN)_2]Br$ crystal structure showing how the 'ET' molecules are arranged in a conducting layer separated by the insulating $Cu[N(CN)_2]Br$ anion layer.

selves into a conducting layer separated by an insulating anion layer (Fig. 10.2). This is strikingly similar to what happens in the layered high T_c cuprates. In that case, we have seen that it is much easier for electrons to move within the copper-oxygen planes than perpendicularly between these planes. There are other similarities between the two families as well in terms of the superconducting members being closely related to magnetic and insulating cousins, and the fact that superconductivity in both families is unconventional and cannot be explained by the standard BCS theory. Organic superconductors are particularly versatile though. By changing the choice of the 'X' molecule, the crystal arrangement and the spacing between the 'ET' molecules can be adjusted and the physical properties of the resulting crystal can be changed. Not only that, but 'ET' is by no means alone as the choice of building block for making these crystals. There is a vast array of possible choices, many of which are flat in shape, but some of which are one-dimensional instead, leading to the exciting possibility of 'designer' superconductors, where things like the spacing between the molecules and the number of electrons that are transferred between the molecules can be tuned at will to dial up the properties you want. However, despite a great deal of effort on the part of organic chemists, the world record T_c of 12 K for this type of organic compound—held since the late 1980s by the compound $(ET)_2$-$Cu[N(CN)_2]Br$—has stubbornly refused to be broken.

This is not the end of the story though. Carbon, because of its extreme versatility, can be assembled into many different structures. One of these is as covalently bonded C_{60} molecules which are usually called *buckyballs* because they have the shape of a ball-shaped cage and were named after *Buckminster Fuller*, the designer of the dome-shaped structures like the famous biomes at the Eden project in Cornwall. If you look at the structure of buckyballs more carefully, they are identical to soccer balls—with

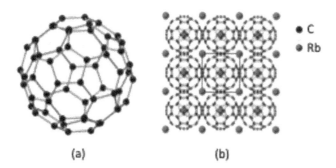

● C
● Rb

(a)   (b)

Fig. 10.3: (Crystal structures of a single buckyball (C_{60} molecule) and Rb_3C_{60} intercalation compound.

carbon atoms arranged around the surface in a network of hexagons and pentagons (see Fig. 10.3(a)). Isolated buckyballs are not much use as a material, but when you have lots of them, they tend to assemble themselves into an ordered arrangement. These crystals are made up of a three-dimensional array of cubic units, with a buckyball placed at the corners and face-centres of each of the cubes in the *face-centred cubic* structure that we met in Section 5.1. Now, this crystal is not superconducting—it is not even metallic—but we can play around with its electronic properties by squeezing in much smaller metal atoms in between the buckyballs. They do not go inside the cage of the C_{60} molecule, but instead go in the spaces in between the molecules (see Fig. 10.3(b)). We call this squeezing in of extra atoms *intercalation* and we make what are known as *intercalation compounds*.[3] For example, alkali metals in group I of the periodic table (like potassium, rubidium or caesium) have only one electron in their outer shell which they are very keen to give away. It turns out that the C_{60} molecule acts as an electron sponge, readily soaking up these electrons. In fact, each buckyball can hoover up as many as 12 extra electrons, but the highest T_c values are achieved if you put in just three alkali metal atoms per C_{60}, thereby donating an extra three electrons to each one.

In terms of the number of electrons donated, it should not matter which of the alkali metals is chosen because they all have one electron to donate. However, it turns out that T_c is quite sensitive to this choice because as you go down the group from potassium (K) to rubidium (Rb) to caesium (Cs), the atoms get physically bigger. If we try to squeeze bigger atoms into the gaps between the buckyballs, there is not quite enough room, so they get pushed apart a little bit and this tends to raise the critical temperature. Chemists have tried to push the buckyballs even further apart using extra intercalated molecules that do not donate any extra electrons themselves, but there comes a point where superconductivity starts to get worse again, usually because it becomes more favourable for the crystal structure to switch from being face-centred

---

[3]The prefix 'inter' means 'between', and intercalation compounds have extra atoms squeezed into the spaces between the existing atoms or molecules in the gaps known as the *interstices*.

cubic to something else and losing superconductivity as a consequence. The highest T_c discovered so far in this family of organic superconductors—the fullerines—is a very respectable 42.5 K in an compound intercalated with a mixture of rubidium and thallium. This compound held the accolade of being the highest T_c superconductor not based on copper oxide until the discovery in 2008 of high T_c superconductivity in the iron-based materials.

Before we leave the organics, there is a final twist to the story—quite literally. In recent times, carbon has reached new heights in terms of its fame, as a result of a couple of scientists in Manchester (UK) discovering that using sticky tape it is possible to break the layered structure of graphite apart repeatedly until you are left with only one single atomic layer. This hexagonally bonded sheet, one atom thick, is known as graphene, and it has been heralded as a wonder material. The hype surrounding it has been a bit like the hype when high T_c superconductors were discovered. As well as being quoted as being 'the strongest known material in the world', electrons can travel about a hundred times faster in it than in silicon (that is used in most of our consumer electronics), it conducts heat twice as well as diamond (the next best material) and even though it is only one atomic layer thick, it is pretty impervious to even the smallest gas molecules like helium. Despite these remarkable properties, it turns out not to be as easy as that to exploit them in real materials and devices. For one thing, although it has a fantastic electrical conductivity, to make electronic devices other properties are important as well as how fast the electrons can move. For example, instead of using copper for our computers, we use silicon even though it has a lower conductivity. The reason is that we can make an electrically controlled switch in silicon—we can turn its conductivity on and off by applying a voltage to a device in the right way. It is not so easy to control the conductivity of graphene in this way, which limits is usefulness in electronics. In fact, there has been a great deal of effort over the past 20 years to try to make graphene behave more like silicon—that is, like a semiconductor rather than a metal.

In a semiconductor, a key feature is that they have something called a *band gap* which is just a gap in energy where there are no electron states. In pure silicon at low temperatures, the electronic states below the band gap are completely filled with electrons and the states above the band gap are completely empty. In this case the material will be an insulator because the lower band is jammed full and the electrons cannot move around to carry charge, and the upper band does not contain any electrons in it at all. We make silicon conducting by putting some electrons into that unoccupied upper band, or removing some electrons from the lower band (to make holes), or both. A single sheet of graphene does not have a band gap which means that we cannot control its conductivity in the same way. However, if we stack two sheets of graphene on top of each other—to make a bilayer—and then apply a voltage across the sheets, it is possible to open up a small band gap. This process is called *gating*. By changing the gate voltage it is possible to tune the band gap, which could be interesting for future device applications.

It is also possible to make bilayer graphene superconduct. Instead of stacking two graphene layers directly on top of each other so that that the carbon hexagons sit directly above one another, if the top layer is twisted slightly relative to the bottom layer we can tune its electronic structure. In particular, there are some special 'magic' twist angles of just over 1° where it is possible to obtain superconductivity when a suitable gate voltage is applied to get the right number of electrons into the system. Although the maximum T_c that has been achieved this way is only 1.7 K—certainly nothing to get excited about in itself—the system is highly tunable making it perfect for studying the fundamental physics of these kinds of quantum materials.

Iron-based superconductors

With the exception of the recently discovered ultra high pressure hydride superconductors, iron-based superconductors are the newest family of superconductors to be discovered. With the current record T_c of over 55 K, they have taken over from the organic superconductors as the second highest T_c superconductors at ambient pressure. Their discovery was particularly unexpected because scientists have always shied away from strongly magnetic atoms like iron when looking for superconductors because it was believed that in any competition between magnetism and superconductivity, magnetism would win. However, it turns out that there is a whole zoo of layered iron-based superconductors, not unlike the array of cuprate HTS materials. In the case of the iron-based materials, the layers that are responsible for superconductivity are usually iron bonded to group V or group VI elements, and they are zig-zag shaped rather in contrast to the copper-oxide layers in the cuprates that are almost flat. The group V elements are known as 'pnictogens' (there is some debate over whether the 'p' is silent or not) and the group VI elements are referred to collectively as 'chalcogens' (cal-co-jens), so you will often hear the iron-based superconductors referred to as 'pnictide' or 'chalcogenide' superconductors. It is possible to stack these iron-pnictide or iron-chalcogenide layers in different ways and put different atoms or molecules in between the layers. The families are often referred to by the ratio of their constituents, in a similar way to the cuprates, and several examples are shown in Fig. 10.4.

Here, we will concentrate first on the highest T_c family which is known as the '1111' or 'eleven-eleven' family because they are based on compounds with the chemical formula LnFeAsO, where Ln is an element in the lanthanide series that appears at the bottom of the periodic table that starts with the element La which has the atomic number of 57. The highest T_c member of this family turns out to be the variant that uses samarium (Sm) as the lanthanide element. If some of the oxygen atoms are

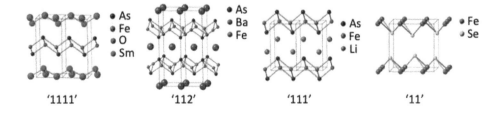

Fig. 10.4: Crystal structures of some iron-based superconductor families.

replaced with fluorine atoms (which have one more electron) it is possible to make this compound superconduct at a temperature of about 55 K. These compounds also have really high upper critical fields of 100–200 tesla, like the cuprates, which make them interesting for high field magnet applications. The main problem with these '1111' oxypnictide superconductors is that they are pretty difficult to make in useful forms. Grain boundaries are weak links, just like they are in cuprates, and so to make wires that can carry high current densities it is probably necessary to use coated conductor technology. It is pretty difficult to grow these materials as thin films though, which is a prerequisite for coated conductors, with the most successful films being grown by a painstakingly slow vacuum deposition process called *molecular beam epitaxy*. The '112' compounds have a lower T_c closer to 30 K, but have the advantage of being considerably easier to make. Grain boundaries do not seem to be quite so catastrophic either which means that cheaper powder-in-tube technology may be an option for making high current wires.

So what about the elephant in the room? It probably has not escaped your notice that these compounds contain arsenic, and arsenic is far from being a friendly element. Like thallium that was discussed in Section 9.1, it is well known to be highly poisonous. It is also volatile, which means that when we heat it up to react it with the other elements, it vaporises into a gas. The processing of these compounds therefore needs to be done very carefully using specialist facilities. There are obvious concerns about scaling up to a safe manufacturing process. An alternative is to use an iron chalcogenide superconductor. The simplest of the chalcogenides is iron selenide which has puckered sheets of FeSe without any additional atoms or molecules pushed in between. The main problem with this compound is that it has a much lower superconducting transition temperature than the pnictides at only around 8 K—not very different to NbTi, our favourite low T_c material. We can push T_c up to about 14 K by replacing half of the selenium atoms with the larger chalcogenide atoms of tellurium. By intercalating alkali metal atoms between the layers of FeSe we can make chalcogenide structures equivalent to the '112' pnictide family, with T_c values of 30–40 K. The problem with these is that alkali metals are very reactive in air, so the superconductors are not stable unless they are kept in an inert environment—not very easy for long lengths of wire. It is also pretty difficult to get them phase pure because they like to separate into two phases, one of which superconducts and the other which is an antiferromagnetic insulator.

FeSe has another trick up its sleeve though. If we grow it as a single monolayer—one FeSe plane thick—on a strontium titanate $SrTiO_3$ crystal, its transition temperature skyrockets to 65 K! In fact, there has been one report where a T_c of 109 K has been achieved—an order of magnitude larger than FeSe in bulk form (T_c= 8 K). Since all iron-based superconductors are unconventional superconductors, it is not surprising that we do not fully understand why this happens. It likely to be partly due to electrons being donated from the strontium titanate substrate, but this in itself is not thought to be sufficient to explain such a large increase in critical temperature. There is other mysterious physics at play.

10.4   Future applications

The upshot of all of this is that, although lots of different superconductors have been discovered since the original high T_c cuprate materials in the late 1980s, none have actually come close to toppling them off their pedestal. The hydrides have higher critical temperature values, but at the cost of needing staggeringly high pressures instead. Although we now understand a great deal more about the physics of superconductivity, we still do not have a definitive theory and we are no closer to having an ambient pressure room temperature superconductor. But do we actually need room temperature superconductors? Is it really the operating temperature that is holding us up in terms of practical deployment of the superconductors that we have already? Well of course, removing all cooling constraints would definitely make life a lot easier and would open up new application areas, but I would argue that even a room temperature superconductor would not see widespread application if it was simply too expensive and difficult to make into a usable high current wire of some sort. The costs we might save by not having to cool it down could easily be overshadowed by the cost of the material itself and unless it has incredible current carrying performance, an expensive wire will never be affordable for mass applications. Before we try to home in on exactly what properties our ideal superconducting material would have, let's look at some of the important new applications that will shape the field over the next few decades.

The next generation of particle accelerators

The Large Hadron Collider (LHC) is a 27 km long circular atom smashing machine buried underground at CERN in France/Switzerland. Essentially it consists of two proton beams—a proton is a type of hadron—that are each accelerated to the enormous energy of 6.5 teraelectronvolts (TeV) where 1 TeV is 10^{12} times as much energy as an electron that has been accelerated by a voltage of 1 volt. Then the protons are forced to collide head on with one another to smash them apart so that the smaller particles produced can be detected. Superconducting magnets are used to steer the proton beams around the tunnel. Scientists use this kind of experiment to try to prove the existence of particles that have been theoretically predicted but not previously detected experimentally, such as the Higgs Boson which was detected for the first time in 2012 at the LHC. The broad aim of these kinds of physics experiments is to unravel the mysteries of the universe, and particle physicists always want to be able to hit these subatomic particles together with higher and higher energy to discover new physics. The next big European project to take over from the LHC is called the *Future Circular Collider*. The idea is that this new machine will reach energies of 100 TeV—around eight times higher than the LHC can achieve at the moment. The problem is that the higher the energy of the protons, the longer the tunnel needs to be unless we are able to increase the strength of the magnets that are used to guide the beam around the circular path. By using HTS magnets operating at 20 tesla rather than Nb_3Sn magnets operating at 16 tesla, the size of the future circular collider could be reduced considerably from 100 km to 80 km circumference. Although this is a single machine, it is so enormous that the shear size of it means that massive quantities of superconductor would be required.

Nuclear fusion power plants

Nuclear fusion is the process by which energy is created in stars like our sun. It involves fusing light atomic nuclei together to make bigger nuclei, releasing huge amounts of energy in the process. If we can harness this process in a special reactor on earth, we would have an almost limitless resource of safe energy. The problem is that enormous temperatures and pressures are required to force the nuclei together so they can fuse. I'm talking about temperatures of around 150 million °C—10 times hotter than the sun! Obviously none of our conventional materials can survive at these kinds of temperatures. Arguably, the most promising way to do it is by using strong magnetic fields to confine the reaction, and that is where superconductors come in.

One of the major problems with making nuclear fusion commercial is that the size of the machines is very large (along with there being huge technical challenges at every step along the way). One way that scientists are planning to make more economical fusion reactors is by changing the shape of the machine to make it more compact (Fig. 10.5). This would require higher magnetic fields that can only be achieved with high temperature superconductors. The compact design also leaves less room for thermal insulation, so it makes sense to use these magnets at temperatures somewhere between 20 K and 35 K which would also preclude low T_c superconductors.

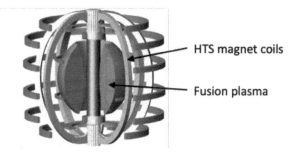

HTS magnet coils

Fusion plasma

Fig. 10.5: Compact spherical tokamak design. (Courtesy of Tokamak Energy.)

Compact reactors bring additional problems for the superconductors though. The reaction that is most likely to be exploited in power plants is the fusion between nuclei of two different isotopes of hydrogen: deuterium (which has a proton and a neutron) and tritium (which has a proton and two neutrons). When they fuse, they form a helium nucleus which has two protons and two neutrons. The extra neutron is emitted with very high energy—14 million electronvolts (14 MeV) to be precise. This means that in a vacuum they would move at around 50,000 kilometres per second. Neutrons are not electrically charged (unlike protons and electrons), and so they do not get slowed down by electrostatic interactions with matter. The only way they are slowed down is when they bang into other nuclei. You may think that you would want to use materials with really heavy nuclei to slow down the fast neutrons, but actually when the light neutrons hit a heavy nucleus they bounce off it with almost the same kinetic energy as they had to start with. However, if they collide with the nuclei of much lighter elements like hydrogen, they can transfer a good fraction of their energy to the other nucleus, slowing them down. So although it might seem counterintuitive, something with lots of light atoms like water or plastic (that are mainly made from carbon and hydrogen atoms) will be better neutron shields than something with heavy atoms. When the neutrons have been slowed down enough, a heavy material like lead or bismuth is useful to capture the neutrons. Basically the neutron fuses with the nucleus when it hits rather than being scattered by it. This in itself might cause further problems down the road because these nuclei with extra neutrons have a tendency to radioactive decay giving out gamma rays.

So what havoc do these high energy neutrons wreak if they reach the superconductor? In simple terms, when a fast neutron hits an atom in the superconductor, it will transfer some of its kinetic energy to the atom which in turn will start to move. This atom will go on to collide with another atom, again transferring some energy and knocking it off its site. This produces a cascade effect and there becomes a zone around the initial impact where many of the atoms have been knocked out of their ideal positions. Thinking about what happens in snooker when the white ball is hit into a pack of reds will give you the right idea. The size of the damaged zone from a neutron collision depends on the energy of the initial neutron and the particular material that it is interacting with. In fact, although a lot of atoms get displaced in one of these collision cascades—the material is often described as melting locally—most of them will find their way back to a sensible position pretty quickly after the initial collision. At least this is what is known to happen in metals that have been studied most intensively. However, it is not surprising that some structural damage remains after such dramatic collision events. In superconductors, these isolated collision cascades actually *improve* the current-carrying performance in the early stages of neutron irradiation. This can be understood by remembering that high critical current densities in type II superconductors arise because of strong pinning of the magnetic flux lines and these extra radiation defects can act as additional flux pinning sites. This is true both in low temperature superconductors like Nb_3Sn that is being used in current generations of fusion reactor, and in high temperature superconductors like (RE)BCO. However, what happens when you irradiate superconductors for too long is that their critical current density starts to plummet, and this happens rather more quickly (at lower neutron doses) in the high T_c materials. This may prove to be rather a headache for compact fusion reactors because the need for high T_c comes about because there is limited space for the magnet windings, and this limited space also means there is not much room for neutron shielding.

Another issue in the design of fusion magnets is that ideally we want to be able to take them apart so that the centre of the reactor can be reached for maintenance purposes. If the magnets are made from a continuous winding like most electromagnets, this would be impossible. Instead the magnets for new fusion reactors will have a modular design—they will be built in separate sections that can be taken apart and put back together by robots. Of course, all of the joints between the cables need to have the minimum possible resistance to avoid them getting too hot. Since low resistance joints are usually made by soldering, the need to be able to take the joints apart and remake them is a major technical challenge. When we are on the subject of joints, in addition to the joints between the sections of cable, within each cable each strand of superconductor needs to be properly connected to enable the current to distribute evenly. Based on simple calculations of how many coated conductor tapes would be needed in a compact fusion reactor, we are looking at there being somewhere in the order of 100,000 tape-to-tape joints. Although we do not need these to have such low resistance as the joints for NMR or MRI magnets that operate in persistent mode without a power supply, every bit of resistance in those joints will add heat load into the system that will need to be taken out with additional cooling power making the whole machine less efficient.

The hydrogen economy

The hydrogen economy is a vision for an environmentally friendly future in which hydrogen will replace fossil fuels as a source of energy. The beauty of using hydrogen as a fuel is that burning it in oxygen only produces water vapour—no greenhouse gases like CO_2 are emitted. It is also an element that is very abundant, although not usually in the form of H_2. One of the ways to make hydrogen is by the electrolysis of water—breaking H_2O molecules apart using electricity produced by renewable sources like wind or solar power. Hydrogen can then be burnt as a fuel to produce thermal energy—a bit like natural gas is used in homes now for cooking and heating water. For vehicles, an alternative is to use fuel cells that are a bit like batteries. They use the reaction of hydrogen with oxygen to produce electricity directly. One of the challenges to realising this cleaner future using hydrogen is the difficulty in storing hydrogen and distributing it to where it is needed. It is such a small molecule that it is hard to contain it in gaseous form because the molecules have a tendency to diffuse out through the container walls. In practice, transporting hydrogen in liquid form—by cooling it down and compressing it—is attractive. In fact, natural gas is typically piped around in liquid form already, so although hydrogen would need to be colder, it is not a completely new technology that would be required.

This idea of transporting hydrogen from A to B as a liquid at cryogenic temperatures opens up the neat possibility of transmitting electricity along superconducting cables in the same pipelines, with the liquid hydrogen acting as the coolant for the superconductor. Liquid hydrogen is a particularly good cryogen, with a low boiling point of 20 K and a better cooling capacity per kilogram than liquid helium or liquid nitrogen. Although it is not cold enough for cheap niobium-titanium low temperature superconductor to be used, it is ideal for the relatively cheap MgB_2 superconductor which works well for this low magnetic field application. Proof-of-concept has already been demonstrated, including the 30 m hybrid energy transfer line in Russia that can transmit up to 135 MW of electrical and chemical power in total.

Unsurprisingly there are challenges to moving towards a hydrogen economy. One of the issues is social acceptance because hydrogen is generally considered to be less safe than other fuels because it can ignite more easily in air. To understand why there is concern you just need to see the footage of the Hindenburg disaster in 1937 in which a Zeppelin (airship), that used huge bags of hydrogen gas to provide the lift, went up in flames as a result of spark of static electricity near a hydrogen gas leak. However, in some respects hydrogen is actually safer than standard fuels—all of which are necessarily flammable—because it is non-toxic and it is so light that it rapidly dissipates if there is a leak instead of building up.

Aircraft of the future

In recent years, air travel has been one of the major contributors to the release of greenhouse gases such as carbon dioxide (CO_2) and nitrogen oxides (NO and NO_2) which are responsible for global warming. For example, figures from just before the COVID 19 pandemic showed that air travel accounted for roughly 2% of all CO_2 emissions induced by human activity worldwide and 12% of transport CO_2 emissions In addition, the release of nitrogen oxides at high altitude are more damaging than release of the same gases at the Earth's surface, adding to the environmental impact of air travel. Ambitious targets for reducing greenhouse gas emissions have been set for the aviation industry, including Europe's Flightpath 2050 initiative which aims to reduce CO_2 emissions by 75% over the first half of the 21st century.

Current aircraft are typically powered by turbofan technology. This has limitations in terms of efficiency because the fan that produces the thrust is mechanically connected to the gas turbine which is powered by burning fuel in the core of the engine. Ideally, to operate at optimum efficiency, the gas turbine rotation speed would be much faster than the fan rotation speed. This limitation could be overcome in theory by separating the two parts of the system and having them connected to each other electrically rather than mechanically. As well as improving the efficiency of each part of the engine, this would open up the design space, allowing many small propulsion fans to be distributed across the wing to optimise the aerodynamics. In this futuristic turbo-electric distributed propulsion technology, the gas turbines would power an electricity generator, which in turn would power electric motors to drive the fans. The problem is that the electrical generators and motors needed to produce the required electrical power are simply too big and too heavy to be feasible for aircraft if they are made from conventional copper or aluminium windings. Conceptual hybrid electric aircraft, such as the NASA N30X hybrid wing body plane rely on using high temperature superconductors with their high current densities because they enable the same power rating in much smaller, lighter weight machines.

Ideally, the biggest savings in terms of power-to-weight ratio would be achieved if high temperature superconductors were used instead of conventional copper wire for all of the components of the motors and generators, as well as for transmitting the power between them. The main challenge, apart from the obvious need for suitable cooling systems, is designing materials and machines that allow AC losses in the superconductor to be minimised. In terms of the cooling, one of the really neat ideas would be to use liquid hydrogen both as the coolant and as the fuel to power the gas turbine!

Superconducting quantum computers

If you have ever run out of space on your mobile phone so that it will not let you take another photo, you will be aware of one of the current limitations with computers—they have finite resources. Information is handled in a computer as a series of zeros and ones: a letter 'A' for example is encoded in your computer as '01000001'. Each one of the digits is a 'bit' of information and the way your computer does calculations is by manipulating the bits using logic gates and outputting a new answer based on the inputs it has received. For example, an 'AND' gate has two inputs. If both inputs are '1' it will output '1', but if either or both of the inputs are '0' it will output '0'. So let's set our computer a simple task that has some relevance to materials science, namely keeping track of the spin state of a system of electrons. If we have just one electron, there are only two possible states it could be in: spin up and spin down. Adding in a second electron doubles the number of possible permutations to four (see Table 10.1). Adding in a third electron doubles the number again to eight. Every time we add an extra electron, we double the number of possible spin states of our system. The number goes up exponentially, and with it the number of bits of information that you need to store. If you have a fairly standard personal computer with 256 gigabytes of storage it would max out at about 40 electrons. The most advanced supercomputers today may be able to handle some 55 electrons—which is still nowhere near what would be needed to model a real material.

Table 10.1 Number of permutations of spin state in a system of electrons.

Number of electrons	Permutations								Number
1	↑	↓							$2^1 = 2$
2	↑↑	↑↓	↓↑	↓↓					$2^2 = 4$
3	↑↑↑	↑↑↓	↑↓↑	↓↑↑	↑↓↓	↓↑↓	↓↓↑	↓↓↓	$2^3 = 8$
n									2^n

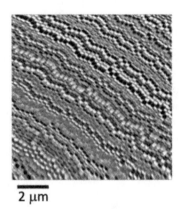

2 μm

Fig. 10.6: Magnetic force microscope image of a magnetic hard disk.

In a classical computer, whether a bit is a '0' or a '1' is controlled by electronic devices called transistors that act a bit like a switch, and if you want to store the bit of information it is written to a special magnetic drive (Fig. 10.6). Each pixel of the magnetic hard drive can be magnetised either up or down to encode the '1' or '0'. Over the decades, the number of transistors that can be fitted on a microchip has doubled about every two years—something known as Moore's law—but there will come a point where it becomes impossible to physically make the individual devices any smaller without quantum effects becoming important. The idea of a quantum computer is to exploit these special quantum mechanical phenomena to make an entirely new kind of computer—a quantum computer—which is based on quantum bits called 'qubits' rather than classical bits.

So what are qubits and how do they differ from classical bits? The main difference is that a classical bit can only take the discrete values of '1' or '0', like a coin that has been flipped can only be 'heads' or 'tails'. In contrast, a qubit can be in both the '1' and the '0' state at the same time because of a special quantum mechanical property called *superposition*. What this means is that if a quantum system can be in a number of different states, those states can be added together and you will end up with another valid quantum state of the system. On a practical level, you can imagine the superposed states as a spinning coin—when it is spinning, it is in some combination of the 'heads' and 'tails' states. It is only when we take a measurement of the system—force the coin to stop spinning—that it becomes either 'heads' or 'tails'. When the system is in a superposition of states, we do not know for certain what state it will end up in when we take a measurement. We only know how likely it is that we will measure a particular state. This may not sound ideal for a computer because you would have thought that this uncertainty in measurement would make it difficult to do calculations, but actually—together with another quantum property called *entanglement*—it enables a whole new way of doing calculations.

Quantum entanglement can also be understood using the coin analogy. Imagine you and I each have an identical coin and we both spin them—put them into a superposition of the 'heads' and 'tails' states—the outcome of my coin (whether I get heads or tails) does not influence the outcome of your coin. However, if the coins were two entangled quantum systems, if I measure my coin first, its outcome will force your coin to have a particular outcome. The coins are no longer in separate quantum states but together in a new entangled quantum state. It may be that it has been entangled in such a way that one coin will read 'heads' and the other will read 'tails', even if the two coins are not physically close to each other.

So how do the properties of superposition and entanglement lead to a different computational approach? Let's say we have three classical bits. Each of them can be in either the '1' or the '0' state, which gives us eight different possible combinations: 000, 100, 010, 001, 110, 101, 011, 111. In a quantum computer with three qubits, each of the qubits can be in a superposition of both '0' and '1' states, so all eight of these combinations exist at the same time. If the qubits are entangled with each other, the computer can essentially test all of these inputs simultaneously instead of one by one.

For every additional qubit you add into the system, the number of permutations doubles. The record number of qubits in a quantum computer available to the public is 53, which corresponds to $2^{53} = 9 \times 10^{15}$ permutations! Now, think back to our example of tracking the spin state of a system of electrons. In the classical computer, we would need to double the number of bits every time we added an extra electron, but in a quantum computer we would only have to add one extra qubit. The computational power of a quantum computer scales exponentially with the number of qubits, and so it is ideally suited to solving problems which also scale exponentially.

But what physical entities can be used as qubits? In fact there are quite a lot of choices, including really cold ions trapped and cooled using lasers, certain atomic defects in diamond, semiconducting quantum dots, nuclear spins and various types of superconducting device. Each of them has its own advantages and disadvantages from a practical point of view. What is needed is a system that can exist in two well defined quantum states, and because superconductors exhibit quantum phenomena even on the macroscopic device scale, they are a particularly practical choice in terms of building a real computer. There are various different types of superconducting qubit which work in different ways. For example, a charge qubit is essentially a superconducting island that contains a certain number of Cooper pairs. We can change the charge by allowing an additional Cooper pair to quantum mechanically tunnel onto the island (which is often known as a Cooper pair box) from a reservoir. It is the different charge—number of Cooper pairs—that gives us the two different quantum states in this case. A flux qubit, on the other hand, is pretty much just like the SQUID device we talked about in Section 6.5. Changing the magnetic flux going through the SQUID loop changes the quantum state. In both types of qubit, Josephson junctions—narrow barriers of non-superconducting material—are essential for their operation. One of the really good things about using superconducting qubits is that we know how to make these devices using standard technologies similar to those used for making integrated circuits for classical computing chips. It is for this reason that the existing quantum computers which you can access right now via the cloud are made using superconducting qubits.

So what are the drawbacks to using superconductors? The main problem is to do with something called *decoherence* and to minimise this effect superconducting qubits have to be cooled to a temperature at least a factor of ten below T_c. When you put a quantum system into a certain state—which is called a coherent state—ideally it would stay in that state indefinitely until we deliberately chose to change it. In practice this does not happen and over time it inevitably loses coherency and it forgets what quantum state it was in. In superconductors, the decoherence time is only about 100 microseconds at best (0.0001 seconds), even when the system is cooled to ultra-low temperatures of less than 0.1 K above absolute zero in a special device called a dilution refrigerator. This does not give very much time for calculations to be performed and means that lots of errors are introduced. In fact, a recent result from a superconducting quantum computer had only 1% signal and the other 99% was noise generated by these errors!

Up until now, physicists have worked on designing cleverer devices to try to slow down decoherence, but as the systems are scaled up to more and more qubits in a real computer, it is essential that the sources of the decoherence are identified and addressed. Pretty much anything that interacts with the qubit will lead to decoherence, and because it is necessary to communicate with the qubits in order to read out their states, it is not possible to isolate them completely. However, we really want to minimise as much unwanted interaction with the environment as possible. One of the ways is by cooling to these extremely low temperatures, but we also have to minimise defects in the materials that might themselves have quantum states that interfere with the qubits. One particular example is the non-superconducting layer that is used as the filling of the superconducting sandwich in a Josephson junction. Typically an aluminium film is used as the superconducting 'bread', and a little bit of oxygen is introduced for a short while part-way through the film deposition process to make an amorphous (non-crystalline) aluminium oxide filling. Unfortunately, it seems as if this aluminium oxide tunnel barrier contains defects that cause unwanted decoherence of the qubits. A limited amount of work has been done so far in trying to grow higher quality tunnel barriers for qubits specifically.

Other materials problems relate to poor surface quality of either the superconductor or the substrate it has been grown on. Aluminium or niobium surfaces are very reactive, which means that any contamination in the environment is likely to stick to the surface and could cause trouble. Nitrides, like titanium nitride or niobium nitride, are also superconductors and it is generally believed that they have cleaner surfaces because they are already saturated with nitrogen—there are no 'dangling bonds' to react with contaminants—so these materials may turn out to be a better choice for quantum computer device applications. Elemental rhenium (Re) is another 'clean' (but very expensive) material that is being investigated for the same reason.

10.5   The 'perfect' superconductor

The question that I am asked most often as a materials scientist working on super-conductors is 'Will we ever have a room temperature superconductor?' Until recently, my answer was always that I have not got a clue. There is nothing special about room temperature from a physics point of view—it is just a temperature about 300 degrees above absolute zero, so there is no particular reason why it is an unbreakable barrier. Of course, recent discoveries have shown that we can certainly get superconductivity at room temperature, but the problem is that the room temperature superconductors that we have got now are not really proper materials. It would be very difficult to engineer anything out of them because they need such high pressures in order to work. Just being able to be in the superconducting state at room temperature is not enough. In fact, accessing low temperatures is not very difficult in the scheme of things, so being able to operate at room temperature is not even close to the top of my personal wish list when it comes to superconducting materials!

Not only that, but the most important properties for our superconductor depend a great deal on what application we want to use the superconductor for. For example, if we want to make an MRI machine, it is vital that our superconductor is cheap. We do not mind too much about how cold we need to operate it because it is a nice enclosed system that is pretty easy to cool with liquid helium. We do not need particularly high magnetic fields for standard MRI, but of course if we could easily increase the field at no extra cost then we might want to do that to improve image quality and decrease scan times (provided it is deemed to be just as safe for the patient). For this reason, the cheapest superconductor, NbTi, is nearly always used. We would only consider switching to a different one if the additional cost in raw material could be offset by the savings in cooling costs or if there is a problem in the future with sourcing niobium.[4] For transmission of electricity, we also do not need to push our material in terms of magnetic field. However, because of the distances that cables extend over, the cooling will be more difficult and expensive. This means that the balance of costs between materials and cooling is flipped. Liquid helium is likely to be too expensive and so it will be more economical to use a high temperature superconductor or possibly MgB_2 that works at intermediate temperatures.

High field magnets have another set of criteria. In this case, materials cost is probably not the be all and end all. What is needed is the highest possible current density at as high a magnetic field as possible. For fusion magnets we would also ideally like to be able to operate at 20 K or above, but for accelerator magnets, 4 K (or even lower) is fine. In applications where the magnetic fields are above 20 tesla, high temperature superconductors are the only choice we have, and we just have to bite the bullet on the materials cost. The conductors used for high field magnets will need to be chock-full of structural defects to make sure that flux pinning is optimised, whereas the materials for transmission cables would probably benefit from fewer defects to reduce flux creep at high operating temperatures and minimise AC loss. Superconducting electronics, like quantum computers, needs another set of properties completely. The superconducting properties that we normally consider to be most important when selecting a superconductor, such as critical temperature, critical field and critical current density, go out the window altogether. Instead it seems to be more important to be able to grow high quality materials with defect-free tunnel barriers and very clean surfaces.

The upshot is that there is no single 'perfect' superconductor. However, if I were to nail my colours to the mast and say what my dream superconductor would be like, top of my wish list would be a ductile material with the mechanical properties and processability of niobium-titanium and the superconducting properties of (RE)BCO. This is a pretty tall order though, not least because in all of the superconductors discovered so far high upper critical field is inexorably linked to a short coherence length. This means that it is unlikely that we will find a high field superconductor in which grain boundaries are not weak links. Even without the discovery of a new wonder material, high temperature superconductors seem to me to be, at last, on the brink of a delivering on their initial promise. Their materials properties are good enough right now, it is just their extremely high cost that is holding them back. If the market pull-through from one of the big applications like fusion comes off, manufacturing costs may be driven down and then high temperature superconductors will, at last, have their day.

---

[4]The estimated resources of niobium are sufficient to meet the worldwide demand in the foreseeable future, but the niobium market is dominated by a few large firms creating a supply risk.

Appendix A

Further reading

Popular books

Ball, P., *Made to Measure: New Materials for the 21st Century*, Princeton University Press (1999).
Blundell, S., *Superconductivity: A Very Short Introduction*, Oxford University Press (2009).
Blundell, S., *Magnetism: A Very Short Introduction*, Oxford University Press (2012).
Ford, P.J. and Saunders, G.A., *The Rise of the Superconductors*, CRC Press (2005).
Gordon, J.E., *The New Science of Strong Materials, or, Why You Don't Fall Through the Floor*, Princeton University Press (2006).
Gordon, J.E., *Structures: or Why Things Don't Fall Down*, Da Capo Press (2003).
Hall, C., *Materials: A Very Short Introduction*, Oxford University Press (2014.)
Miodownik, M., *Stuff Matters*, Penguin (2013).
Polkinghorne, J., *Quantum Theory: A Very Short Introduction*, Oxford University Press (2002).
Street and Alexander, *Metals in the Service of Man*, Penguin (1944).

Primers

Ashby, M.F. and Jones, D.R.F., *Engineering Materials 2*, 4th edition, Elsevier Ltd (2013).
Atkins, P.W., *Physical Chemistry*, 9th edition, W.H. Freeman (2010).
Callister and Rethwisch, *Materials Science and Engineering*, 9th edition, Wiley (2014).
Cottrell, A., *An Introduction to Metallurgy*, 2nd edition, Routledge, 1997.
Dove, M.T., *Structure and Dynamics, An atomic view of materials*, Oxford University Press (2003).
Feynman, R.P., Leighton, R.B. and Sands, M.L., *The Feynman Lectures on physics*, Volume II, Definitive edition, Pearson/Addison-Wesley (2006).
Hull, D. and Bacon, D.J., *Introduction to dislocations*, 5th edition, Elsevier Ltd (2011).
Jones, D.R.F. and Ashby, M.F., *Engineering Materials 1*, 5th edition, Elsevier Ltd (2018).
Porter, D.A., Easterling, K.E., Sherif, M.Y., *Phase Transformations in Metals and Alloys*, 3rd edition, CRC Press (2018).
Sutton, A.P., *Concepts of Materials Science*, Oxford University Press (2021).

Specialist books

Annett, J.F., *Superconductivity, Superfluids and Condensates*, Oxford University Press (2004).

Blundell, S., *Magnetism in Condensed Matter*, Oxford University Press (2001).

Cardwell, D.A. and Ginley, D.S., (eds), *Handbook of Superconducting Materials*, IOP Publishing Ltd (2003).

Narlikar, A.V., *Superconductors*, Oxford University Press (2014). Rogalla, H. and Kes, P.H. (eds), *100 years of superconductivity*, CRC Press (2012).

Singleton, J., *Band Theory and Electronic Properties of Solids*, Oxford University Press (2001). Wilson, M.N., *Superconducting Magnets*, Oxford University Press (1987).

Appendix B

Field decay and joint resistance

A superconducting coil in persistent mode operation can be modelled by the circuit diagram in Fig. B.1. You will probably have come across electrical circuits with resistors in them before. Here the coil has a property other than its resistance that affects the voltage across it called its *inductance* (L). This is shown as the spiral element called an *inductor* in the circuit diagram.

The magnetic field in a solenoid coil is directly proportional to the current that is flowing in the wire. We can define a quantity called the *magnetic flux* (ϕ) which is just the amount of magnetic field that passes through a particular cross-sectional area in space. For a loop of wire, we are interested in the flux going through the loop, so the flux will be given by the magnetic field (B) multiplied by the cross-sectional area of the loop. For a solenoid that has many turns of wire (lots of loops of the same area), we work out the flux through each single loop and add them all up to get the total flux. You may have come across the idea that B is more properly called the *magnetic flux density*[1] rather than the more generic *magnetic field*. This is because it is the flux per unit area and it distinguishes it from another kind of magnetic field (called H).

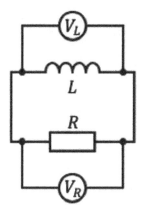

Fig. B.1: Circuit diagram of a superconducting magnet with a joint.

[1]It is sometimes even called the *magnetic induction*.

In this book we will not use H fields at all so I will refer to B as simply the magnetic field or the magnetic field strength.

We define inductance (L) of a coil as the constant of proportionality between the total magnetic flux through the coil and the current in the wire that produces the magnetic field in the first place: $\phi = LI$. But what is the physics behind the voltage that is dropped across an inductor? Well, if you have come across Faraday's law, you will have already learnt that if the magnetic field through a coil changes with time, you generate an electromotive force (voltage). Lenz's law tells us that the voltage that is produced is in the correct sense to try to oppose the change in field. This means that if the field decreases slightly, a voltage will be generated to increase the current flowing through the coil to try to boost the field again. We can write Faraday's law for the voltage across an inductor (V_L) neatly using calculus.

$$V_L = -\frac{d\phi}{dt} = -L\frac{dI}{dt} \tag{B.1}$$

The circuit also has a small resistance as well as the inductance. The superconducting wires themselves have zero resistance, but there is likely to be a small resistance introduced by the joints in the circuit, however carefully we make them. We can see from the circuit diagram that because we are operating without a power supply, the voltage across the inductor must be equal to the voltage across the resistor.

$$V_L = V_R \tag{B.2}$$

$$-L\frac{dI}{dt} = IR \tag{B.3}$$

$$\frac{dI}{dt} = -\frac{R}{L}I \tag{B.4}$$

This is a type of equation called a *first-order differential equation* and they are incredibly common in applied physics of all sorts (as well as in things like rates of population growth of bacteria colonies). The standard method for solving this type of equation involves putting all of the terms containing I on one side of the equation and all of the terms including t on the other side, and integrating each side separately. We call this *separating the variables*.

$$\frac{dI}{dt} = -\frac{R}{L}I \tag{B.5}$$

$$\frac{1}{I}dI = -\frac{R}{L}dt \tag{B.6}$$

$$\int_{I_0}^{I(t)} \frac{1}{I}dI = -\frac{R}{L}\int_0^t dt \tag{B.7}$$

$$\ln\left(\frac{I(t)}{I_0}\right) = -\frac{Rt}{L} \tag{B.8}$$

$$I(t) = I_0 e^{-\frac{Rt}{L}} \tag{B.9}$$

$$I(t) = I_0 e^{-\frac{t}{\tau}} \tag{B.10}$$

This analysis shows that the current through the circuit decays exponentially with time. How quickly it decays is characterised by the time constant $\tau = \frac{L}{R}$ which tells us the time it takes for the current to drop to $\frac{1}{e}$ of its original value. The larger the joint resistance, the more quickly the current through the coil (and hence the magnetic field) will decay.

For a typical MRI magnet, we need a very high field stability—the magnetic field (current) must change by less than 0.1 part per million per hour (<0.1 ppm hr^{-1}). This is the same as saying that the field dropped in an hour has to be less than 10^{-7} of the original field.

$$\frac{B_0 - B(1hr)}{B_0} = 1 - \frac{B(1hr)}{B_0} = 10^{-7} \tag{B.11}$$

A typical value of the inductance of an MRI magnet is 20 henry (H),[2] so we can estimate how much resistance we can get away with in the magnet by rearranging equation B.10.

$$R = -\frac{L}{t} \ln\left(\frac{B(t)}{B_0}\right) \tag{B.12}$$

$$R = -\frac{20}{60 \times 60} \ln(1 - 10^{-7}) \tag{B.13}$$

$$R \approx 10^{-10} \, \Omega \tag{B.14}$$

Since there are about 20 joints in an MRI magnet, each one has to have a resistance $<5 \times 10^{-12} \, \Omega$. In reality MRI manufacturers aim for an order of magnitude less than this which means their magnets have decay time constants as high as 10,000 years!

---

[2]The henry is the SI unit of inductance and is equivalent to 1 kg m^2 s^{-2} A^{-2}.

Appendix C

Derivation of condensation energy

Imagine we change the conditions of our system (e.g. temperature, pressure, magnetic field) so that the internal energy changes from U_i (initial internal energy) to U_f (final internal energy), we define the change in internal energy as $\Delta U = U_f - U_i$. ΔU will be positive if the internal energy increases, and negative if the internal energy decreases. The 'Δ' symbol represents the change in a property when we change the conditions and can be used with any of the state functions—not just internal energy. ΔU is also used to describe the difference in the internal energy between two phases (α and β) under the same conditions: $\Delta U = U^\alpha - U^\beta$.

However, we often want to know what happens to our state functions when we make an infinitesimally small change in conditions. We use the notation 'dU' to refer to an infinitesimally small change in the internal energy. You may have come across the use of symbol d before in the context of differentiation. The first derivative of y with respect to x is usually denoted $\frac{dy}{dx}$ and it effectively tells us how sensitive to change parameter y is to an infinitesimal change in x. When we just refer to dU we are not specifying what we have done to make this infinitesimal change in U—we might have changed the temperature slightly, or we might have changed the applied magnetic field slightly, or we might have done both. It is, therefore, a rather general notation.

We are going to start deriving the condensation energy in a superconductor from the first law of thermodynamics:

$$\Delta U = Q + W \tag{C.1}$$

where Q is the heat energy and W is the work done on the system by the surroundings. We can write this in terms of infinitesimal changes in heat and work as $dU = dQ + dW$. The heat energy transferred to the system $dQ = TdS$, where dS is the change in its level of disorder (entropy). dW is the magnetic work done by the magnetic field B to change the magnetisation (magnetic moment per unit volume) of the material by an infinitesimal amount dM, and is given by $dW = VBdM$, where V is the volume of the system.[1]

---

[1] In general, we would also need to include what is called 'PV' work—the mechanical work associated with changing the volume of a gas at pressure P but we can ignore this as we do not have a gas in our system.

Substituting these expressions into the first law of thermodynamics gives an equation for the infinitesimal change in internal energy that comes about by putting in an infinitesimal amount of heat and/or doing an infinitesimal amount of magnetic work on the system.

$$dU = TdS + VBdM \tag{C.2}$$

When we are talking about a magnetic material, we define enthalpy (H) by subtracting the total magnetic work done to magnetise the material from the internal energy of the system, giving $H = U - VBM$. By substituting this into the definition of Gibb's free energy, we find

$$G = H - TS = U - VBM - TS \tag{C.3}$$

Now we do a trick: we are really interested in how Gibb's free energy *changes* when we change conditions like temperature and magnetic field rather than what its absolute value is, so we differentiate this equation. However, we need to differentiate every term with respect to every parameter that may change. The only thing that we can consider to be a constant in our case is the volume of the material. Everything else needs to be differentiated. To do this we use a mathematical technique called *partial differentiation*. Here I will just give the result, but the eagle-eyed of you may notice that you need to use the 'product rule' to do it.

$$dG = dU - VBdM - VMdB - TdS - SdT \tag{C.4}$$

Substituting our the expression for dU in equation C.2 into this equation we find that the TdS and $VBdM$ terms cancel and we arrive at a fairly simple equation for the change in free energy as a result of infinitesimal changes in magnetic field and temperature.

$$dG = (TdS + VBdM) - VBdM - VMdB - TdS - SdT \tag{C.5}$$
$$= -VMdB - SdT \tag{C.6}$$

To work out an expression for the condensation energy we are going to use the situation shown in Fig. C.1. Here we start with our superconductor at temperature T and zero field (point P). We are interested in calculating the free energy change that occurs when we increase the field from zero to the critical field B_c (going from point P to point Q on the diagram). Because temperature is not changing $dT = 0$ and equation C.6 becomes

$$dG = -VMdB \tag{C.7}$$

So now we need to know how the magnetisation M in a superconductor relates to the applied field B. Because superconductors are perfect diamagnets in the Meissner state and they push out all of the applied magnetic field, we would expect its magnetisation to scale with field and have the opposite sign. It turns out that $M = -\frac{B}{\mu_0}$ where μ_0 is the magnetic permeability of free space (vacuum) and is a constant ($\mu_0 = 4\pi \times 10^{-7}$ Hm^{-1}). The presence of the constant μ_0 is an inconvenience that comes about because

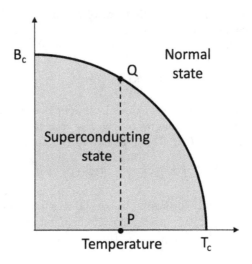

Fig. C.1: Schematic diagram of the magnetic phase diagram of a superconductor.

of the historical definitions of B and M that we do not need to concern ourselves with here. It just means that this extra factor comes into all of our magnetic equations. By substituting into equation C.7 we can write $dG = \frac{VB}{\mu_0} dB$.

The total free energy change when we increase the applied field from point P to point Q is therefore given by equation C.9 where the 'S' superscripts refer to the fact that the sample is in the superconducting state.

$$\Delta G^S = G^S(Q) - G^S(P) \tag{C.8}$$

$$= \frac{V}{\mu_0} \int_0^{B_c} B dB = \frac{VB_c^2}{2\mu_0} \tag{C.9}$$

We could imagine doing exactly the same calculation for the hypothetical case where the material is in the normal state instead of the superconductor. In this case the sample does not have to expel the magnetic field that is applied at Q because in the normal state the material is not a perfect diamagnet anymore. We assume it is completely non-magnetic and so we find

$$\Delta G^N = G^N(Q) - G^N(P) = 0 \tag{C.10}$$

But how does this help us work out the condensation energy for the superconductor at this temperature? Condensation energy is the amount of energy saved by going from the normal state to the superconducting state at zero applied field (i.e. at point P on the diagram). Therefore what we are trying to calculate is $G^S(P) - G^N(P)$. To do this we need to consider one more part of the puzzle. We know that the superconducting and normal states must be in equilibrium with each other at point Q when $B = B_c$. This means that the free energy in the superconducting and normal states must be the same at point Q $(G^S(Q) = G^N(Q))$. Putting this together with equations C.9 and C.10 we arrive at an expression for the condensation energy in terms of the critical field (a property that can be measured experimentally).

$$\text{Condensation energy} = G^S(P) - G^N(P) \tag{C.11}$$

$$= (G^S(Q) - \Delta G^S) - (G^N(Q) - \Delta G^N) \tag{C.12}$$

$$= (G^S(Q) - G^N(Q)) - \frac{V B_c^2}{2\mu_0} \tag{C.13}$$

$$= -\frac{V B_c^2}{2\mu_0} \tag{C.14}$$

List of Symbols and Abbreviations

$2\Delta_0$	superconducting energy gap.
Φ	magnetic flux.
Φ_0	magnetic flux quantum.
ψ	wave function.
σ	stress.
ε	strain.
ξ	coherence length.
Ω	ohm: unit of resistance.
μ_0	permeability of free space.
ρ	resistivity.
AC	alternating current.
AEC	alkaline earth cuprate.
amp	ampere: unit of electrical current.
APT	Atom probe tomography.
atm	atmosphere: unit of pressure.
B	magnetic field.
b	Burgers vector.
B_c	critical field.
BCC	body-centred cubic.
B_{c1}	lower critical field.
BCS	Bardeen–Cooper–Schrieffer theory.
B_{c2}	upper critical field.
Bi-2212	$Bi_2Sr_2CaCu_2O_8$.

Bi-2223	$Bi_2Sr_2Ca_2Cu_3O_{10}$.
BSCCO	Bi-Sr-Ca-Cu-O.
buckyball	Buckminster fullerine (C_{60}) molecule.
CERN	European Council for Nuclear Research.
D	diffusion coefficient.
DC	direct current.
E	Young's modulus.
ET	BEDT-TTF (Bis(ethylenedithio)tetrathiafulvalene) molecule.
eV	electronvolt.
F	force.
FCC	face-centred cubic.
E_F	Fermi energy.
G	Gibb's free energy.
H	henry: unit of inductance.
H	enthalpy.
h	Planck's constant.
HCP	hexagonal close-packed.
HTS	high temperature superconductor.
HVDC	high voltage direct current.
I	current.
I_c	critical current.
ITER	International nuclear fusion tokamak.
J	joules.
J	current density.
J_c	critical current density.
J_e	engineering critical current density.

K	kelvin: unit of absolute temperature.
k	spring constant.
k_B	Boltzmann's constant.
L	inductance.
LHC	Large Hadron Collider.
LTS	low temperature superconductor.
M	magnetisation.
MgB_2	magnesium diboride.
MRI	magnetic resonance imaging.
N	Newton: unit of force.
Nb_3Sn	niobium-tin.
NbTi	niobium-titanium.
nm	nanometre.
NMR	nuclear magnetic resonance.
P	power.
Pa	pascal: unit of pressure.
qubit	quantum bit.
RABiTS	rolling assisted biaxially textured Substrate.
RE	rare-earth element.
RE-123	$REBa_2Cu_3O_7$.
RE-211	$RE_2Ba_2CuO_5$.
(RE)BCO	$REBa_2Cu_3O_7$.
R	resistance.
RF	radio frequency wave.
RRP®	Resacked Rod Process.
S	entropy.
SEM	scanning electron microscope.

SMES   superconducting magnetic energy storage.

SQUID   superconducting quantum interference device.

STM   scanning tunnelling microscope.

T   tesla: unit of magnetic field.

T_c critical temperature.

TEM   transmission electron microscope.

U internal energy.

V   volt: unit of voltage.

Wb   weber: unit of magnetic flux.

wt %   weight percent.

YBCO $YBa_2Cu_3O_7$.

References

Bednorz, J.G. and Müller, K.A. (1986). Possible high T_c superconductivity in the Ba La Cu O system. *Z. Physik B–Condensed Matter*, **64**, 189–193.

Blundell, Stephen (2009). *Superconductivity: A Very Short Introduction*. OUP.

Boutboul, T., Le Naour, S., Leroy, D., Oberli, L., and Previtali, V. (2006). Critical current density in superconducting Nb Ti strands in the 100 mt to 11 t applied field range. *IEEE Transactions on Applied Superconductivity*, **16**(2), 1184–1187.

Braccini, V, Xu, A, Jaroszynski, J, Xin, Y, Larbalestier, D C, Chen, Y, Carota, G, Dackow, J, Kesgin, I, Yao, Y, Guevara, A, Shi, T, and Selvamanickam, V (2010, dec). Properties of recent IBAD–MOCVD coated conductors relevant to their high field, low temperature magnet use. *Superconductor Science and Technology*, **24**(3), 035001.

Cosmus, Thomas C. and Parizh, Michael (2011). Advances in whole-body mri magnets. *IEEE Transactions on Applied Superconductivity*, **21**(3), 2104–2109.

Dimos, D., Chaudhari, P., Mannhart, J., and LeGoues, F. K. (1988, Jul). Orientation dependence of grain-boundary critical currents in $\mathbf{Yba_2cu_3o_{7-\delta}}$ bicrystals. *Phys. Rev. Lett.*, **61**, 219–222.

Drozdov, A., Eremets, M., Troyan, I., Ksenofontov, V., and Shylin, S.I. (2015). Conventional superconductivity at 203 kelvin at high pressures in the sulfur hydride system. *Nature*, **525**, 73–76.

Duperray, G. and Herrmann, P.F. (2003). Processing of high T_c conductors: the compound Bi(2212). In *Handbook of Superconducting Materials, Volume I* (ed. D. Cardwell and D. Ginley), pp. 449–477. Institute of Physics.

Eastell, C. J., Moore, J. C., Fox, S., Everett, J., A.D., Caplin., Grovenor, C.R.M., and Goringe, M.J. (1998). Tem investigation of the microstructure and properties of tl(bi)-1223/ag powder-in-tube superconducting tapes. *Philosophical Magazine A*, **78**(1), 57–84.

Ekino, Toshikazu, Takasaki, Tomoaki, Ribeiro, Raquel, Muranaka, Takahiro, and Akimitsu, Jun (2007, mar). Scanning tunnelling microscopy and spectroscopy of MgB2. *Journal of Physics: Conference Series*, **61**, 278–282.

Geilhufe, R. Matthias, Borysov, Stanislav S., Kalpakchi, Dmytro, and Balatsky, Alexander V. (2018, Feb). Towards novel organic high-T_c superconductors: Data mining using density of states similarity search. *Phys. Rev. Materials*, **2**, 024802.

Herrmann, C and Rock, A (2012). Magnetic resonance equipment (mri)–study on the potential for environmental improvement by the aspect of energy efficiency. Technical report, not known.

Jiang, Jianyi, Bradford, Griffin, Hossain, S. Imam, Brown, Michael D., Cooper, Jonathan, Miller, Evan, Huang, Yibing, Miao, Hanping, Parrell, Jeff A., White, Marvis, Hunt, Andrew, Sengupta, Suvankar, Revur, Rao, Shen, Tengming, Kametani, Fumitake, Trociewitz, Ulf P., Hellstrom, Eric E., and Larbalestier, David C. (2019). High-performance bi-2212 round wires made with recent powders. *IEEE Transactions on Applied Superconductivity*, **29**(5), 1–5.

Kamihara, Y., Hiramatsu, H., Hirano, M., Kawamura, R., Yanagi, H., Kamiya, T., and Hosono, H. (2006). Iron-based layered superconductor: LaOFeP. *J. Am. Chem. Soc.*, **128**, 10012–10013.

Kanithi, H., Blasiak, D., Lajewski, J., Berriaud, C., Vedrine, P., and Gilgrass, G. (2014). Production results of 11.75 Tesla Iseult/INUMAC MRI conductor at Luvata. *IEEE Transactions on Applied Superconductivity*, **24**(3), 1–4.

Koch, C.C. and Easton, D.S. (1977). A review of mechanical behaviour and stress effects in hard superconductors. *Cryogenics*, **17**(7), 391–413.

Larbalestier, D., Jiang, J., Trociewitz, U., Kametani, F., Scheuerlein, C., Dalban-Canassy, M., Matras, M., Chen, P., Craig, N. C., Lee, P. J., and Hellstrom, E. E. (2014). Isotropic round-wire multifilament cuprate superconductor for generation of magnetic fields above 30 t. *Nature Mater*, **13**, 375–381.

Lee, P.J. (1999). Abridged metallurgy of ductile alloy superconductors. In *Wiley Encyclopedia of Electrical and Electronics Engineering, Vol. 21* (ed. J. Webster), pp. 75–87. New York: Wiley.

Li, G.Z., Sumption, M.D., Zwayer, J.B., Susner, M.A., Rindfleisch, M.A., Thong, C.J., Tomsic, M.J., and Collings, E.W. (2013, jul). Effects of carbon concentration and filament number on advanced internal Mg infiltration-processed MgB_2 strands. *Superconductor Science and Technology*, **26**(9), 095007.

Majewski, Peter (2000, 04). Materials aspects of the high-temperature superconductors in the system Bi_2O_3-SrO-CaO-CuO. *Journal of Materials Research*, **15**, 854 – 870.

Nagamatsu, J., Nakagawa, N., Muranaka, T., Zenitani, Y., and Akimitsu, J. (2001). Superconductivity at 39 k in magnesium diboride. *Nature*, **410**, 63–64.

Namburi, D.K., Shi, Y., and Cardwell, D.A. (2021, jan). The processing and properties of bulk REBCO high temperature superconductors" current status and future perspectives. *Superconductor Science and Technology*, **34**, 053002.

Parrell, J.A., Field, M.B., Zhang, Y., and Hong, S. (2004). Nb_3Sn conductor development for fusion and particle accelerator applications. *AIP Conference Proceedings*, **711**(1), 369–375.

Parrell, J.A., Zhang, Y., Field, M.B., Cisek, P., and Hong, S. (2003). High field Nb_3Sn conductor development at Oxford Superconducting Technology. *IEEE Transactions on Applied Superconductivity*, **13**(2), 3470–3473.

Rogalla, H. and Kes, P.H. (2012). *100 years of superconductivity*. CRC Press.

Sanabria, C., Field, M., Lee, P.J., Miao, H., Parrell, J., and Larbalestier, J.C. (2018, apr). Controlling cu–sn mixing so as to enable higher critical current densities in RRP nb3sn wires. *Superconductor Science and Technology*, **31**(6), 064001.

Snider, E., Dasenbrock-Gammon, N., McBride, R., Debessai, M., Vindana, H., Vencatasamy, K., Lawler, K.V., Salama, A., and Dias, R.P. (2020). Room-temperature superconductivity in a carbonaceous sulfur hydride. *Nature*, **586**, 373–377.

Song, X. Braccini, V. and Larbalestier, D.C. (2004). Inter- and intragranular nanostructure and possible spinodal decomposition in low-resistivity bulk mgb2 with varying critical fields. *Journal of Materials Research*, **19**, 2245–2255.

Swift, R.M. and White, D. (1957). Low temperature heat capacities of magnesium diboride (MgB$_2$) and magnesium tetraboride (MgB$_4$). *J. Am. Chem. Soc.*, **79**, 3641–3644.

Tarantini, C., Kametani, F., Balachandran, S., Heald, S.M., Wheatley, L., Grovenor, C.R.M., Moody, M.P., Su, Y-F., P.J., Lee, and Larbalestier, D.C. (2021). Origin of the enhanced Nb$_3$Sn performance by combined Hf and Ta doping. *Sci Rep*, **11**, 17845.

Thomas, Heiko, Marian, Adela, Chervyakov, Alexander, Stückrad, Stefan, Salmieri, Delia, and Rubbia, Carlo (2016). Superconducting transmission lines – sustainable electric energy transfer with higher public acceptance? *Renewable and Sustainable Energy Reviews*, **55**, 59–72.

Index

The manufacturer's authorised representative in the EU for product safety is
Oxford University Press España S.A. of el Parque Empresarial San Fernando
de Henares, Avenida de Castilla, 2 – 28830 Madrid (www.oup.es/en).

Printed in the USA/Agawam, MA
December 13, 2024

878851.023